AF750828

Health Policy and Law in Japan

Jiro Ezaki • Shoichi Maeda • Taisuke Izumi
Editors

Health Policy and Law in Japan

Springer

Editors
Jiro Ezaki
Ministry of Health, Labour and Welfare,
The Government of Japan
Chiyoda-ku, Tokyo, Japan

Department of Biology, College of
Arts & Sciences
American University
Washington, DC, USA

Taisuke Izumi
Department of Biology, College of
Arts & Sciences
American University
Washington, DC, USA

Shoichi Maeda
Graduate School of Health Management,
Division of Health Care Management
and Public Health
Keio University
Shinjuku-ku, Tokyo, Japan

ISBN 978-981-95-9240-1 ISBN 978-981-95-9241-8 (eBook)
https://doi.org/10.1007/978-981-95-9241-8

© The Editor(s) (if applicable) and The Author(s), under exclusive license to Springer Nature Singapore Pte Ltd. 2026

This work is subject to copyright. All rights are solely and exclusively licensed by the Publisher, whether the whole or part of the material is concerned, specifically the rights of translation, reprinting, reuse of illustrations, recitation, broadcasting, reproduction on microfilms or in any other physical way, and transmission or information storage and retrieval, electronic adaptation, computer software, or by similar or dissimilar methodology now known or hereafter developed.
The use of general descriptive names, registered names, trademarks, service marks, etc. in this publication does not imply, even in the absence of a specific statement, that such names are exempt from the relevant protective laws and regulations and therefore free for general use.
The publisher, the authors and the editors are safe to assume that the advice and information in this book are believed to be true and accurate at the date of publication. Neither the publisher nor the authors or the editors give a warranty, expressed or implied, with respect to the material contained herein or for any errors or omissions that may have been made. The publisher remains neutral with regard to jurisdictional claims in published maps and institutional affiliations.

This Springer imprint is published by the registered company Springer Nature Singapore Pte Ltd.
The registered company address is: 152 Beach Road, #21-01/04 Gateway East, Singapore 189721, Singapore

If disposing of this product, please recycle the paper.

Preface

The Jefferson Memorial, as seen across the tidal basin during peak cherry-blossom bloom. Japan gifted the cherry trees along the Potomac River in Washington, DC, in 1912 as a symbol of peace and friendship between Japan and the United States. More than 100 years later, these cherry trees delight many citizens

As a public health physician working for the Japanese government, I had the opportunity to study abroad in the United States for two years, from 2023 to 2025. I researched forensic medicine, medical law, and health policy in the United States at the University of Maryland, Baltimore, and American University (Washington, DC).

While in the United States, I frequently received questions from faculty members and students: "I hear that Japan has the highest average life expectancy and lowest infant mortality rate in the world. What measures are being taken?" "How does the universal health insurance system function?" and "How will Japan overcome the challenges of an aging society?" These questions highlighted the significant interest in Japan's health policies and laws. At the same time, I realized that most of the materials on these topics were written in Japanese, and no introductory books or English textbooks explained the system's overall picture. I received an offer from Springer Nature to publish a book, and together with Dr. Izumi of American

University and Dr. Maeda of Keio University, I decided to write this book, *Health Policy and Law in Japan*.

The year 2025 marks the 80th anniversary of the end of World War II. Through postwar reconstruction and rapid economic growth, Japan's social security system has become well developed, and the average life expectancy and infant mortality rate of Japanese people are among the best in the world.

In this book, I trace Japan's 80-year history of health policy since the end of World War II. It explains how Japan's health policy and legal system were established and developed into their current form, as well as their basic concepts and characteristics. This book is intended as an introductory text for students and researchers majoring in medicine or public health at US universities. However, it is also an ideal first-hand resource for public health officials seeking to understand Japan's systems and pharmaceutical company executives considering Japan as a market.

This book will contribute to a better understanding of Japan's health policy and, ultimately, to the health of humanity.

At the laboratory at American University, Washington, DC

Jiro Ezaki

Contents

Editors and Contributors

About the Editors

Jiro Ezaki received his MD and PhD from the School of Medicine, Hiroshima University, Japan. After graduating from Hiroshima University, he completed a clinical residency program at the Tokyo Women's University Hospital. In 2014, he became an Assistant Professor at the Department of Legal Medicine, Tokyo Women's Medical University, Japan. As a forensic pathologist, he focused on research involving traumatic brain injury (TBI), drug overdose, and Medical Law.

Since 2016, he has served as a public health physician, Deputy Director, at the Ministry of Health, Labour, and Welfare, the Government of Japan. He worked on government public health policies involving patient safety, medico-legal death investigations, and rare and intractable diseases.

From 2019 to 2020, he was Director of the Public Health Division in Kochi Prefecture, responsible for managing prefecture-wide public health measures. He worked on cancer policy (resident screening and palliative care), infectious disease control (COVID-19, Hepatitis Virus, and STD), and maternal and child health promotion (Child Fatality Review, Child Abuse, and Infant Health Checkups).

In addition to his posts in the government, he contributed to public health and medical law education for medical and nursing students in Japan by giving lectures, training, and writing textbooks as a Visiting Associate Professor at Hiroshima University, the Invited Faculty at Osaka University, and the Senior Research Fellow at the SFC Institute of Keio University. In 2025, the Government of Japan dispatched him to the Izumi Lab, Department of Biology, College of Arts and

Sciences, American University, to engage in education and research on Japanese health policy and law.

Shoichi Maeda, PhD, is the dean of the Graduate School of Health Management and a Professor in the Division of Health Care Management and Public Health at Keio University. He also holds a joint appointment as a professor in the Department of Health Policy and Management at the Keio University School of Medicine. After earning his PhD in Medical Sciences from Kyushu University in 2001, he held faculty positions at the Graduate School of Medical Sciences at Kyushu University and the University of Tokyo Graduate School of Medicine before joining Keio University in 2009.

His research focuses on medical law, medical ethics, public health ethics, research ethics, and healthcare risk management. He has published widely on topics including medical errors, informed consent, end-of-life care, and the management of incidental findings in clinical practice and medical research. He has also studied US legislation concerning medical assistance in dying (MAID).

He has authored numerous books and research papers, contributed to educational resources, and helped develop Japanese government guidelines on ethical issues in life sciences and medical research. He has also served on ethics and conflict-of-interest committees for leading Japanese medical societies.

In addition to his work in Japan, he currently serves as a faculty fellow at American University and leads collaborative research projects with multiple institutions worldwide.

Taisuke Izumi, PhD, is Assistant Professor of Biology at American University, where his research focuses on viral persistence, host-virus interactions, and the development of functional cures for chronic infections such as HIV. He earned his PhD in medical science from Kyoto University, Japan, and completed postdoctoral training at the National Institutes of Health, MD, USA.

He has published over 40 peer-reviewed articles and secured multiple research grants, including NIH R15 support and a DC CFAR Pilot Award. His laboratory has developed innovative strategies for HIV and SARS-CoV-2 research in collaboration with academic, federal, and industry partners.

He has guided students at the undergraduate, master's, and doctoral levels, many of whom have earned competitive awards, publications, and placements in graduate and professional programs. He is also committed to inclusive teaching, developing active learning curricula, and building partnerships that connect biology with entrepreneurship, regulatory science, and industry.

In addition to his scholarship and teaching, he is involved in various departmental, collegiate, and university committees, including those for master's and PhD programs and student awards and fellowships. He further contributes to professional service as an editor for multiple journals and as a member of NIH grant review committees. Furthermore, he actively participates in outreach programs, such as the SMART high school research internship. His efforts are enhancing biomedical research at AU and promoting interdisciplinary collaborations across the sciences.

Contributors

Maeve Engelbride Department of Biology, College of Arts & Sciences, American University, Washington, DC, USA

Jiro Ezaki, MD, PhD Ministry of Health, Labour and Welfare, The Government of Japan, Chiyoda-ku, Tokyo, Japan
Department of Biology, College of Arts & Sciences, American University, Washington, DC, USA

Sophia Gagliardi Department of Biology, College of Arts & Sciences, American University, Washington, DC, USA

Tristan Hotchkin Department of Biology, College of Arts & Sciences, American University, Washington, DC, USA

Yuka Iijima, MBBS Department of International Health, Johns Hopkins Bloomberg School of Public Health, Baltimore, MD, USA

Eri Ishikawa, MHS Graduate School of Health Management, Division of Health Care Management and Public Health, Keio University, Shinjuku-ku, Tokyo, Japan

Jun Izawa, PhD Japan Medical Association, Bunkyo-ku, Tokyo, Japan

Coco Izumi Department of Biology, College of Arts & Sciences, American University, Washington, DC, USA

Taisuke Izumi Department of Biology, College of Arts & Sciences, American University, Washington, DC, USA

Mei Kobayashi Department of Public Health and Health Systems, Graduate School of Medicine, Nagoya University, Nagoya, Japan

Shoichi Maeda Graduate School of Health Management, Division of Health Care Management and Public Health, Keio University, Shinjuku-ku, Tokyo, Japan

Daniel Mansour, PharmD, AGSF, BCGP, FASCP The Peter Lamy Center on Drug Therapy and Aging, Department of Practice, Science, and Health Outcomes Research, University of Maryland, School of Pharmacy, Baltimore, Maryland, USA

Dacia Marquez Department of Biology, College of Arts & Sciences, American University, Washington, DC, USA

Hasset Tibebe Department of Biology, College of Arts & Sciences, American University, Washington, DC, USA

Takashi Yamashita, PhD Department of Sociology, Anthropology, and Public Health, College of Arts, Humanities, and Social Sciences, University of Maryland, Baltimore County, Baltimore, MD, USA

Hiroshi Yatsuya, MD, PhD Department of Public Health and Health Systems, Graduate School of Medicine, Nagoya University, Nagoya, Japan

Ayaan Zafar University of Illinois College of Medicine Chicago, Chicago, IL, USA

Invitation to Japan

1

Jiro Ezaki, Yuka Iijima, Shoichi Maeda, and Taisuke Izumi

Before reviewing the progress of the health policy in Japan in the past 80 years, this chapter describes what kind of country Japan is, including its geography, climate, population, and anthropological characteristics.

1.1 The Japanese Archipelago

The Japanese archipelago is an arc-shaped chain of islands in the northwestern Pacific Ocean, located off the eastern coast of the Eurasian continent in East Asia. Its widest point is approximately 3500 km long and 300 km wide. It has four main islands—Hokkaido, Honshu, Shikoku, and Kyushu—and over 14,000 smaller islands. When we impose the map of the Japanese archipelago on the US East Coast, its total length extends from Maine to Florida (Fig. 1.1) [1].

J. Ezaki
Ministry of Health, Labour and Welfare, The Government of Japan, Chiyoda-ku, Tokyo, Japan

Department of Biology, College of Arts & Sciences, American University, Washington, DC, USA

Y. Iijima
Department of International Health, Johns Hopkins Bloomberg School of Public Health, Baltimore, MD, USA

S. Maeda
Graduate School of Health Management, Division of Health Care Management and Public Health, Keio University, Shinjuku-ku, Tokyo, Japan

T. Izumi (✉)
Department of Biology, College of Arts & Sciences, American University, Washington, DC, USA
e-mail: izumi@american.edu

© The Author(s), under exclusive license to Springer Nature Singapore Pte Ltd. 2026

J. Ezaki et al. (eds.), *Health Policy and Law in Japan*,
https://doi.org/10.1007/978-981-95-9241-8_1

Fig. 1.1 Japan superimposed at the same latitude on the US. When we impose the map of the Japanese archipelago on the US East Coast, its total length extends from Maine to Florida [1]

Japan's climate varies greatly, ranging from the subarctic environment of Hokkaido in the north to the subtropical marine climate of Okinawa in the south, which we can easily imagine considering the environment of Maine and Florida in the US. Most of the archipelago is characterized by warm and rain-rich conditions, with the land being eroded and steep due to the influence of the rainy season "Tsuyu," which spans from May to early July, typhoons in September, seasonal winds, and heavy snowfall [1]. Japan's distinct four seasons shape not only its natural landscapes but also its cultural identity—delicate cherry blossoms herald the arrival of spring (Fig. 1.2), shimmering coastlines and festivals enliven the summer, vibrant foliage blankets the mountains in autumn, and serene snowfall transforms northern regions into quiet, white vistas in winter.

Fig. 1.2 Cherry blossoms in Tokyo, Japan. Cherry blossoms (Sakura), Chidorigafuchi moat, Tokyo Imperial Palace. https://photock.jp/photo/chidorigafuchi-17

Mountains cover approximately 75% of Japan's land area [1]. Since ancient times, the Japanese have lived on limited flat land, cultivating rice as their staple food and developing a unique food culture and lifestyle. Understanding Japanese food culture and lifestyle is crucial for comprehending Japan's disease structure and preventing lifestyle-related diseases.

1.2 Japanese People: Where Did They Come From?

Humanity is believed to have first appeared on Earth around 5 million years ago. During the Ice Age, the Japanese archipelago was physically connected to the Asian continent, allowing people and animals to move freely across the land. As the climate warmed and sea levels rose approximately 10,000 years ago, Japan gradually became isolated as a group of islands. Fossils of extinct elephants such as Palaeoloxodon naumanni and Mammuthus primigenius—found in ancient geological layers across Japan—provide compelling evidence of this ancient land bridge to the mainland (Fig. 1.3) [2].

The origin of the Japanese people has been a long-standing theme in Japanese anthropological society. In 1991, Japanese biological anthropologist Dr. Kazuo Haniwara proposed the theory that the Japanese people were formed through the intermarriage of two groups: the Jomon people, who primarily lived by hunting and gathering, and immigrants, namely Torai people or Yayoi people, who brought rice cultivation to the Japanese archipelago and made it their main livelihood. Today's Japanese history textbooks also introduce the theory that people descended from the Southern Mongoloid lineage from southern Asia formed the prototype of the Japanese people, and that, later, people from the Northern Mongoloid lineage (immigrants) from northern China migrated to Japan and repeatedly intermarried with the Jomon people who were already there, leading to the formation of the Japanese people [2]. Additionally, the Ainu people in Hokkaido and the Ryukyuan people of the Ryukyu Islands still retain strong physical characteristics of ancient Mongoloids [2].

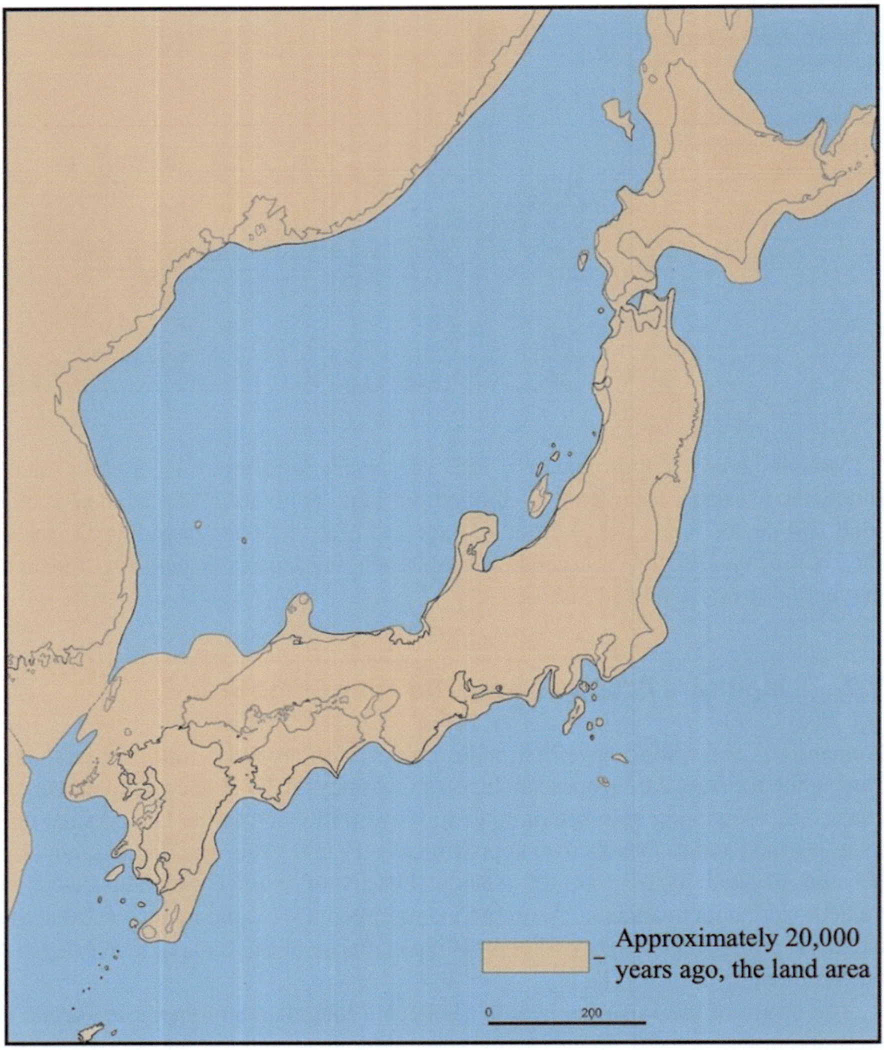

Fig. 1.3 The Japanese archipelago during the Last Ice Age (20,000 years ago). During the Ice Age, Japan is thought to have been connected to the Eurasian continent by land https://www.mogurin.or.jp/museum/hwm/details/tenzi01/t01_1_japan.html

Recent genomic studies linking ancestral origins to disease patterns have revealed intriguing connections between Japan's anthropological history and modern public health challenges. For example, the "Human T-cell Leukemia Virus Type I" (HTLV-1), which causes leukemia, lymphoma, and myelopathy, was discovered in 1973 by Japanese hematologist Dr. Yorifumi Hinuma (1925–2015) [3]. There are 10 million HTLV-1 virus carriers globally, with approximately 1 million in Japan, and preventing vertical transmission through breast milk is a critical public health issue. In

Japan, it is particularly prevalent in Kyushu, the Ryukyu Islands, and Hokkaido, with a higher carrier rate in regions where many people inherit traits from the Jomon people. A recent anthropological study on genomic analysis identified genomic variant mutations involving Jomon people. Interestingly, the study showed that areas with more Jomon-derived genetic variations are more prone to obesity, whereas regions with a lower proportion are more prone to severe asthma [4].

1.3 Population

Japan's population in 2025 is approximately 120 million, making it one of 14 countries with a population exceeding 100 million. With a population of 340 million, the US is roughly one-third the size of Japan. Considering that Japan's land area is one-twenty-sixth that of the United States, Japan has an extremely high population density compared to the US.

Looking back at long-term population trends, Japan's population has grown rapidly over the past 150 years. The population was between 5 and 7 million in the seventh century, reaching 10 million in the fifteenth century and 30 million in the seventeenth century. After World War II, Japan's population reached 80 million in 1948 and peaked at 128 million in 2008. Now, it is declining (Fig. 1.4) [5].

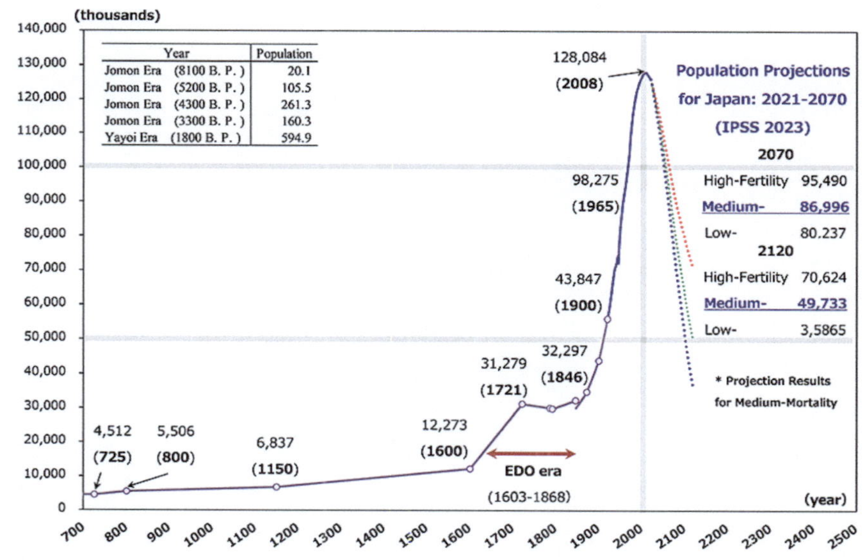

Source: Statistics Bureau, Ministry of Internal Affairs and Communications, Population Census of Japan. Unless otherwise stated in the notes, the table is concerned with the population of the whole national area (homeland) that the administrative authority has covered as of October 1 of each year. Therefore, the data for 1945-1970 do not include Okinawa Prefecture. : National Institute of Population and Social Security Research, Population Projections for Japan: 2021-2070 (2023). The projected population as of October 1 of each year. Hiroshi Kito, Jinkou Kara Yomu Nihon No Rekishi (Kodan-sha, 2000). Yuzo Morita, Jinko Zouka No Bunseki (Nippon-hyoron-sha, 1944).

Fig. 1.4 Long-term Population Trends in Japan

Japan has entered an era of declining birth rates and rapid population aging. While many high-income countries, including Japan, face this demographic challenge, Japan's population structure has several unique characteristics.

- Life expectancy at birth is among the highest in the world, at 81.05 years for men and 87.09 years for women (as of 2022).
- Due to rapid aging, the proportion of the elderly population is the highest globally and continues to increase annually (in 2023, the proportion of those aged 65 and older accounted for 29.1% of the total population).
- In recent years, the population has been declining due to a decrease in births and an increase in deaths (in 2023, the total population was 124.35 million, a decrease of 600,000 from the previous year).
- The population is heavily concentrated in urban areas.

The age structure of the population is represented by a diagram called a "population pyramid (Fig. 1.5)" [1]. However, the population structure of our country, which is in such a situation, is no longer pyramid-shaped but rather vase-shaped. Population and population structure changes influence social security systems, economic conditions, and disease patterns, leading to changes in attitudes toward illness and health and people's work and lifestyle cycles.

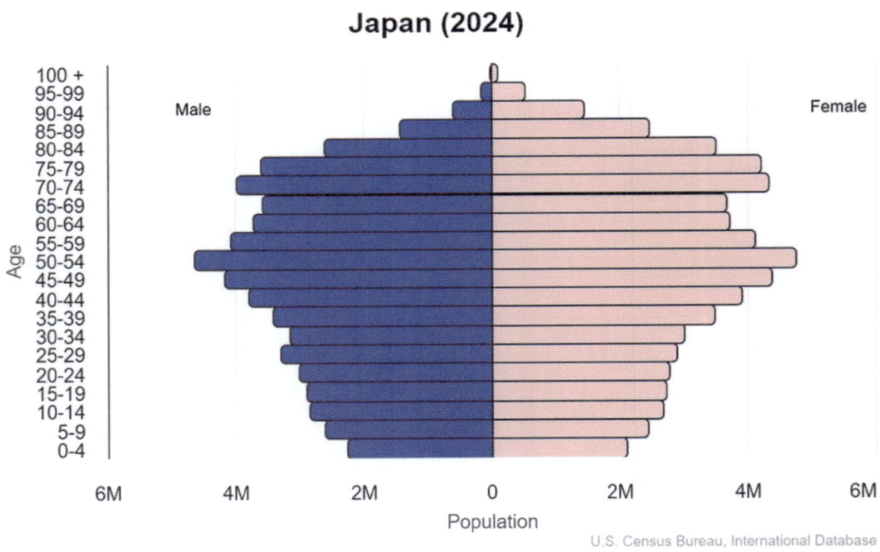

Fig. 1.5 Population Pyramid (Japan, 2024). Japan's population pyramid is "vase-shaped"

References

1. The World Factbook 2024. Washington, DC: Central Intelligence Agency, 2024. https://www.cia.gov/the-world-factbook/countries/japan/#people-and-society.
2. Shin Mouichidoyomu Yamakawa Nihonnshi (Japanese History for High School, in Japanese), Yamakawa Shuppansha, ISBN-13: 978–4634590908.
3. Otani M, et al. Phylogeography of human T-lymphotropic virus type 1 (HTLV-1) lineages endemic to Japan. Trop Med Health. 2012;40(4):117–24.
4. Watanabe Y, Ohashi J. Modern Japanese ancestry-derived variants reveal the formation process of the current Japanese regional gradations. iScience. 2023;26(3):106130.
5. Kousei-roudou hakusho (2015) (in Japanse) (Ministry of Health, Labour and Welfare's White Paper), ISBN-13: 978–4865790351, data available from URL: https://www.mhlw.go.jp/wp/hakusyo/kousei/15/backdata/01-00-01-001.html

History of Medicine

2

Yuka Iijima, Shoichi Maeda, Taisuke Izumi, and Jiro Ezaki

In this chapter, we introduce medicine and public health before and during World War II (1945). Furthermore, we will look back on how Japan reached the starting point for building its social security system after World War II.

2.1 Edo Period (1603–1868)

In Japanese history, the period from 1603 to 1868 is known as the Edo period. During this time, the country was ruled by the Tokugawa shogunate, with the central government based in Edo, now Tokyo [1].

The Edo period was marked by approximately 260 years of peace and stability [1]. Economic growth and social order allowed samurai, merchants, and

Supplementary Information The online version contains supplementary material available at https://doi.org/10.1007/978-981-95-9241-8_2.

Y. Iijima
Department of International Health, Johns Hopkins Bloomberg School of Public Health, Baltimore, MD, USA

S. Maeda
Graduate School of Health Management, Division of Health Care Management and Public Health, Keio University, Shinjuku-ku, Tokyo, Japan

T. Izumi
Department of Biology, College of Arts & Sciences, American University, Washington, DC, USA

J. Ezaki (✉)
Ministry of Health, Labour and Welfare, The Government of Japan, Chiyoda-ku, Tokyo, Japan

Department of Biology, College of Arts & Sciences, American University, Washington, DC, USA
e-mail: jezaki@american.edu

© The Author(s), under exclusive license to Springer Nature Singapore Pte Ltd. 2026

J. Ezaki et al. (eds.), *Health Policy and Law in Japan*,
https://doi.org/10.1007/978-981-95-9241-8_2

townspeople to pursue education and cultural activities. An isolationist foreign policy, known as sakoku ("closed country"), restricted contact with the outside world and contributed to the development of distinctly Japanese cultural traditions, for example, the tea ceremony called "Sadō" (Fig. 2.1), flower arrangement, and sumo wrestling.

Medicine and public health in the Edo period centered on "Kampo medicine," a traditional Japanese medicine based on Chinese medicine. Doctors made house calls to treat patients at home and prescribed medications (see column "Story of Kampo" in this chapter). At that time, the concept of admitting patients to hospitals for treatment was not established, so hospitals did not exist. An exception was the "Koishikawa Yojosho"(currently known as the Koishikawa Botanical Garden in Bunkyo Ward, Tokyo), established in 1722 by Tokugawa Yoshimune, the eighth Shogun.

It is believed that Koishikawa Yojosho housed and cared for poor patients, who received medicine made from herbs harvested in the garden.

One of the most famous physicians of the Edo period was Hanaoka Seishu. In 1804, Hanaoka successfully performed breast cancer surgery using general anesthesia. This is recognized as the first documented case of general anesthesia using herbal anesthetics in a surgical procedure, preceding the use of ether anesthesia in the West by over 40 years (Fig. 2.2).

Kampo medicine did not include surgery, and under the isolationist policy, only those who had studied Dutch medicine by translating Dutch medical books attempted surgical procedures. The first Western-style hospital in Japan was established in Nagasaki in 1861 by Dutch naval surgeon Pompe, who had been sent to Japan despite the isolationist policy. Pompe and his successor, Baldwin, operated the hospital and provided medical education to many Japanese people at the Nagasaki Health Center. Many Japanese who studied medicine under them contributed to the development of modern medicine in Japan.

Fig. 2.1 Japanese traditional tea ceremony "Sadō". Serving Matcha tea to guests using traditional methods

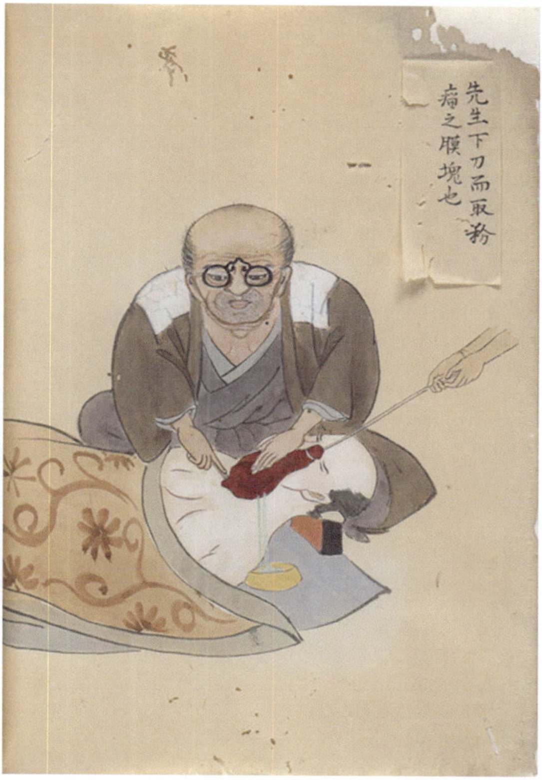

Fig. 2.2 Hanaoka Seishu (1760–1835) and Anesthesia. (From Nihon iji bunka iryoushuusei, vol 2, Nihon Ishigakkai)

2.2 Modernizing Japan (Meiji, Taisho Era) and World War II (1853–1945)

In 1853, four US Navy warships (the "Kurofune," the Black ship in Japanese) led by Commodore Matthew Perry (Fig. 2.3) arrived on the coast of Uraga, Kanagawa prefecture (Fig. 2.4) [1]. In 1854, the Japan-U.S. Treaty of Peace and Amity was signed, opening Japanese ports to American vessels. However, the Edo period officially came to an end in 1867, when the Tokugawa shogunate relinquished power to the emperor in an event known as the "Taisei Hōkan" (Restoration of Imperial Rule) [1]. Furthermore, in 1867, the Edo shogunate surrendered its authority to Emperor Meiji, marking the end of the Edo period. With this, the Meiji government was established, Edo was renamed "Tokyo," and Japan began transforming into a modern Western-style nation.

The Meiji government adopted Western-style government structures such as the Cabinet, the National Diet (Parliament), and the courts. It announced its policy of adopting Western medicine for public health administration. The Medical Bureau was established within the Ministry of Education of the Japanese government in

Fig. 2.3 Commodore Matthew Perry (1809–1855). Matthew Calbraith Perry (1794–1858) was a United States Navy officer who commanded ships in several wars, including the War of 1812 and the Mexican–American War. He led the Perry Expedition, which ended Japan's isolationism, and signed the Convention of Kanagawa between Japan and the United States in 1854

Fig. 2.4 Kurofune in Uraga, Kanagawa (1853)

1872 [2]. In 1874, Japan's first modern medical and public health regulations, the Isei, were enacted. The Isei aimed to "protect the people's health, treat diseases, and promote the study of medicine." Its promulgation marked the first step toward the modernization of public health administration.

Under the Isei, while the prefectural government had the authority to permit the opening of hospitals, there were no nationwide specific permission criteria [2]. Moreover, there were no rules involving smaller medical facilities (clinics). The former Medical Practitioners Act was enacted in 1906, and a registration system for physicians to open clinics was established, significantly contributing to Japan's modern medical system. Meanwhile, the number of hospitals and clinics increased, and cases of hospitals being established by non-physicians and for-profit companies grew; issues such as the commodification of medical care and the spread of false or exaggerated advertising arose. Therefore, the central government started restricting hospital and clinic advertising through a ministerial ordinance in 1909 [2]. Subsequently, in 1933, the government revised the former Medical Practitioners Act, creating legal regulations under central government authority for the first time. Non-physicians required permission from the governor to establish hospitals and clinics, and detailed rules regarding the management structure and facilities of hospitals and clinics were established under the "Clinic Regulation Rules"(1933) [2, 3]. These regulatory frameworks were maintained through World War II and inherited by the Medical Care Act in the present (see Chap. 5). These regulations were under the jurisdiction of the Ministry of the Interior's Public Health Bureau until the establishment of the Ministry of Health and Welfare (the predecessor of the Ministry of Health, Labour and Welfare) in 1938 [3].

With the Second Sino-Japanese War outbreak in 1937, medicine and public health policies also shifted to a wartime system. During this period, public health measures aimed to reduce tuberculosis and infant mortality rates, increase the population, and improve people's physical strength for national defense [3].

The Health Center Act, enacted in 1937, established health centers as governmental organizations that provide health guidance and consultation for the public. Health centers, namely "Hokenjo" in Japanese, were defined as organizations responsible for providing necessary health guidance at the local level to improve the physical condition of the population. By the 1937 fiscal year, 49 health centers had been established nationwide, with an additional 187 centers established over the next five years. After this, at its peak, more than 800 health centers will be established nationwide [3].

In 1938, the Ministry of Health and Welfare was established as a separate entity from the Ministry of the Interior to promote "improvement of national physical strength" and "enhancement of national welfare." Subsequently, matters related to health centers, including public health administration, came under the jurisdiction of the Ministry of Health and Welfare's Health Bureau [3].

In 1942, the authority to issue instructions regarding physical fitness improvement and orders for medical treatment, which had previously been under the jurisdiction of local governors, was transferred to health center directors. As a result, health centers assumed the dual role of advisory bodies and administrative agencies responsible for implementing measures [3].

2.3 Post-War Japan (1945–Present)

On August 15, 1945, Japan surrendered unconditionally to the Allied Powers and was placed under occupation. Under the leadership of the Supreme Commander for the Allied Powers (SCAP, General Headquarters, GHQ), the country underwent demilitarization and democratization, and the wartime system was completely dismantled (Fig. 2.5) [1].

The Constitution of Japan was enacted in November 1946. The constitution provides for a parliamentary system and three branches of government: the National

Fig. 2.5 General MacArthur lands at Atsugi Airfield (Kanagawa, Japan) (1945)

Diet (legislative), the Cabinet led by a Prime Minister (executive), and the Supreme Court (judicial), which are the highest bodies of power. It guarantees individual rights, including legal equality, freedom of assembly, association, speech, due process, and a fair trial. Article 9 of the Constitution renounces Japan's right to wage war and to maintain military forces. The Constitution of Japan has not been amended once since its enactment in 1947.

Article 25 states "All people shall have the right to maintain the minimum standards of wholesome and cultured living." The article also mentions "In all spheres of life, the State shall use its endeavors for the promotion and extension of social welfare and security, and public health." The Constitution stipulates the basic principles of Japan's social security system, which is significant for Japan's recovery and economic growth [3, 4].

Additionally, in May 1946, based on the GHQ memorandum "Regarding the Reform of the Health and Welfare Administrative System," three new bureaus—the Public Health Bureau, the Medical Affairs Bureau, and the Preventive Medicine Bureau—were established within the Ministry of Health and Welfare in the same year. In December, health policy bureaus were established within local governments (currently prefectures).

Under this system, in April 1947, the GHQ issued a memorandum titled "The Expansion and Strengthening of the Health Center System." In response, the Health Center Act was completely revised in September of the same year and came into effect in January 1948. With this revision, public health police duties such as food hygiene and prevention of acute infectious diseases, which police agencies had previously handled, were transferred to the jurisdiction of health centers, strengthening their role as frontline institutions for public health (Video 2.1).

GHQ/SCAP occupied Japan for only seven years, from 1945 to 1952, during which time most of today's fundamentals of Japan's public health were established. measures against infectious disease (1902–1994), a U.S. Army physician and the Civil Affairs Occupation of Japan, Commander, United Nations Public Health &

Welfare Detachment, contributed to improving public health in Japan by strengthening health centers, implementing school lunch programs, implementing measures against infectious diseases, and separating pharmaceuticals from medical care.

In 1950, the Social Security System Council, an advisory committee to the government, issued recommendations on the social security system based on Article 25 of the Constitution, outlining the following general principles:

- Article 25 of the Constitution of Japan clearly states that all citizens have the right to life and that the state must ensure their livelihood.
- The so-called "social security system"refers to measures to provide economic security through insurance or direct public expenditure to address the causes of poverty, to ensure the minimum standard of living for those who have fallen into poverty through national assistance, and to promote public health and social welfare, thereby enabling all citizens to lead lives worthy of members of a cultural society.
- Since the state assumes responsibility, citizens must also fulfill the social obligations necessary for maintaining and operating this system by their respective abilities, based on the spirit of social solidarity.
- The core of social security shall be a social insurance system in which individuals bear the necessary expenses. At the same time, the state shall directly assist those in need who cannot be rescued by the insurance system alone and guarantee their minimum standard of living. Furthermore, governmental efforts must simultaneously promote public health measures to maintain and improve people's health and expand social welfare administration to prevent the breakdown of national life.

The recommendation was truly visionary. It outlined the direction in which Japan would take over the next 80 years in building its social security system [4]. This book introduces how the social security system was developed over the 80 years after WW2 through Health policy and Law.

2.4 Column "Story of Kampo Medicine"

Kampo medicine (Kanpō igaku), literally meaning the "method from Han period (206 BC to 220 AD) of ancient China," represents Japan's distinctive traditional medical system, predominantly based on herbal remedies. Introduced from China via the Korean Peninsula around the fifth to sixth CEnturies CE, early Kampo closely adhered to Chinese classical medicine, heavily influenced by herbal therapies, foundational theories, and canonical texts. However, through extensive adaptation influenced by Japan's unique flora, climate, and sociocultural dynamics, Kampo evolved into a distinctively Japanese form of medicine [5].

The Edo period (1603–1867) was a pivotal epoch for Kampo, marking its establishment as the primary medical practice throughout Japan. During this era, the emergence of influential schools such as the Koho-ha ("Ancient

Method") significantly shaped Kampo's theoretical and practical landscape. The Koho-ha practitioners emphasized rigorous adherence to classical Han-dynasty Chinese medical texts, notably the "Shang Han Lun" and the "Jin Gui Yao Lue," distinguishing Kampo from the contemporaneous evolution of Traditional Chinese Medicine (TCM), which underwent distinct transformations in mainland China.

Kampo medicine notably diverges from modern TCM through several distinctive practices. Predominantly, Kampo employs standardized herbal formulations instead of the individualized herb combinations commonly utilized in TCM. A hallmark of Kampo diagnosis is "Fukushin," an abdominal palpation technique uniquely refined in Japan. This meticulous diagnostic method involves palpating specific regions of the abdomen to detect subtle abnormalities such as tension, tenderness, and pulsation, thereby providing critical insights into the patient's internal physiological states.

The Meiji Restoration in 1868 posed significant challenges to Kampo medicine as Japan rapidly embraced Western medical paradigms, relegating traditional methods to secondary status. Unique among East Asian nations, Japan adopted a unified medical licensure system exclusively focused on Western medical education. Consequently, Kampo was preserved as a complementary practice among Western-trained physicians rather than as a standalone medical discipline. Despite initial setbacks, the mid-twentieth century witnessed a pronounced resurgence of Kampo, driven by Japanese physicians advocating for the scientific validation and integration of traditional medicine within modern clinical frameworks.

In contemporary Japan, Kampo is thoroughly integrated into the healthcare system, officially recognized under Japan's National Health Insurance since 1967. Currently, 148 standardized Kampo prescriptions are reimbursable, ensuring stringent quality control measures for safety and efficacy [6]. Empirical data demonstrate that over 80% of Japanese physicians prescribe Kampo regularly in clinical practice, particularly within gynecology, where its usage approaches universal adoption for managing women's health issues such as menstrual disorders and menopausal symptoms [7]. Furthermore, formal medical education in Japan now mandates foundational training in Kampo, underscoring its recognized clinical significance.

Central to Kampo practice is the concept of "Sho," representing a holistic diagnostic pattern of physiological imbalance rather than merely a disease-specific diagnosis. Kampo practitioners employ comprehensive diagnostic methods, including visual inspection, pulse diagnosis, tongue examination, and notably the abdominal palpation (Fukushin), to ascertain the patient's Sho (Fig. 2.6). Accurate determination of Sho directs the precise selection of herbal formulations from an established compendium, facilitating individualized yet standardized therapeutic approaches.

The historical trajectory of Kampo—from its initial roots in classical Chinese medicine to its sophisticated development as a distinctly Japanese therapeutic system—illustrates a compelling model for harmonizing traditional medical insights with contemporary scientific medicine. Its successful integration within Japan's healthcare infrastructure not only underscores its practical utility but also exemplifies the potential for traditional medical systems to contribute meaningfully alongside modern biomedical practices.

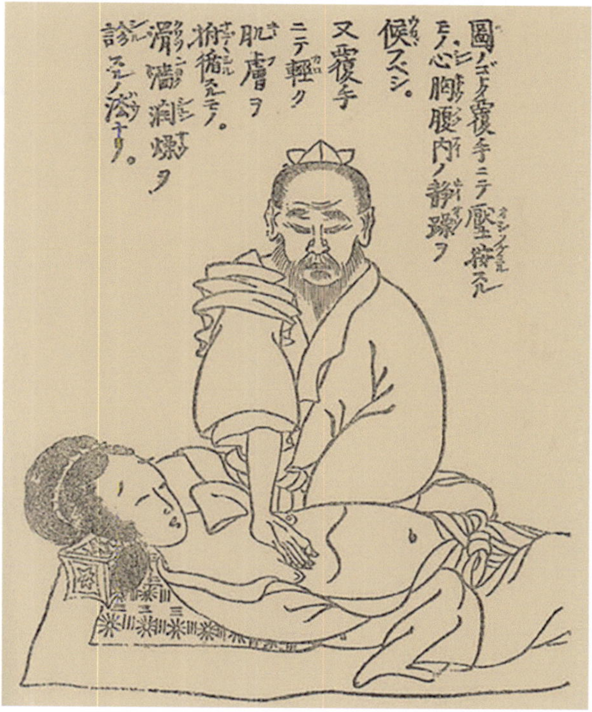

Fig. 2.6 Illustrations of Classical Fukushin Method. (From Fukusho Kiran Yoku, 1809)

References

1. Shin Mouichidoyomu Yamakawa Nihonnshi (Japanese History for High School, in Japanese), Yamakawa Shuppansha, ISBN-13: 978-4634590908.
2. Nihon no iryoseido no tokucho ha sono rekishi kara umareta (no.3), Kousei no shihyo. 2016 Sep; 63 (11) 50–54 (In Japanese).
3. Kousei-roudou hakusho (2014) (in Japanse) (Ministry of Health, Labour and Welfare's White Paper), p. 4–42, ISBN-13: 978-4905427872.
4. Kousei-roudou hakusho (2017) (in Japanse) (Ministry of Health, Labour and Welfare's White Paper), p. 1–32, ISBN-13: 978-4-86579-104-4.
5. Sreedhar R, Watanabe K, Arumugam S. Chapter 1 – introduction to Japanese Kampo medicines. In: Arumugam S, Watanabe K, editors. Japanese Kampo medicines for the treatment of common diseases: focus on inflammation [internet]. Academic Press; 2017. p. 1–11. [cited 2025 May 10]. Available from: https://www.sciencedirect.com/science/article/pii/B9780128093986000019.
6. Watanabe K, Matsuura K, Gao P, Hottenbacher L, Tokunaga H, Nishimura K, et al. Traditional Japanese Kampo medicine: clinical research between modernity and traditional medicine-the state of research and methodological suggestions for the future. Evid Based Complement Alternat Med. 2011;2011:513842.
7. Moschik EC, Mercado C, Yoshino T, Matsuura K, Watanabe K. Usage and attitudes of physicians in Japan concerning traditional Japanese medicine (Kampo medicine): a descriptive evaluation of a representative questionnaire-based survey. Evid Based Complement Alternat Med. 2012;2012:139818.

Medical Professions

3

Jiro Ezaki, Daniel Mansour, Shoichi Maeda,
and Taisuke Izumi

In Japan, a Medical Doctor (M.D.) is referred to as a "Medical practitioner" in medical law. The Medical Practitioners Act was enacted three years after the promulgation of the Constitution of Japan [1]. According to Article 1, "Medical practitioners shall contribute to the improvement and promotion of public health by providing medical care and health guidance, thereby ensuring the healthy lives of the people." [1]. As can be seen from using the phrase "improvement and promotion of public health" in Article 25 of the Constitution (see Chap. 2), medical practitioners are expected to approach their duties with a mindset that respects human life and dignity.

They are expected to treat illnesses and injuries and take the lead in improving and promoting public health in all fields. Medical practitioners must be professionals in both clinical medicine and public health [1]. Regardless of whether they engage in clinical practice, physicians are expected to strive always to prevent diseases among the people and promote their health.

J. Ezaki
Ministry of Health, Labour and Welfare, The Government of Japan,
Chiyoda-ku, Tokyo, Japan

Department of Biology, College of Arts & Sciences, American University, Washington, DC, USA

D. Mansour
The Peter Lamy Center on Drug Therapy and Aging, Department of Practice,
Science, and Health Outcomes Research, University of Maryland, School of Pharmacy,
Baltimore, Maryland, USA

S. Maeda
Graduate School of Health Management, Division of Health Care Management and Public
Health, Keio University, Shinjuku-ku, Tokyo, Japan

T. Izumi (✉)
Department of Biology, College of Arts & Sciences, American University,
Washington, DC, USA
e-mail: izumi@american.edu

© The Author(s), under exclusive license to Springer Nature Singapore Pte Ltd. 2026
J. Ezaki et al. (eds.), *Health Policy and Law in Japan*,
https://doi.org/10.1007/978-981-95-9241-8_3

Similar expressions can be found in the Dental Practitioners Act for Doctor of Dental Surgery (DDS, dentists), the Pharmacists Act for Doctor of Pharmacy (Pharm.D.), which are based on the concept that the goal of all providers of medical care is to improve public health [2, 3]. Dentists are defined in Article 1 of the Dental Practitioners Act as "Dental practitioners are to contribute to the improvement and promotion of public health through the administration of dental care and health guidance, and thereby ensure the healthy lives of the citizens." [2]. Pharmacists are defined in Article 1 of the Pharmacist Act as "those who contribute to the improvement and promotion of public health by dispensing medications, supplying pharmaceuticals, and overseeing pharmaceutical hygiene, thereby ensuring the health of the public." [3]. In comparison, pharmacists in the United States have seen their roles expand significantly over the last 30 years, particularly in relation to ensuring medication safety and optimizing patient outcomes.

3.1 Principles

If you were a governmental public health physician responsible for formulating medical policies, what regulations would you establish to ensure that medical care is provided appropriately?

One approach would be for the government to enact laws such as "When treating Disease A, physicians must administer Drug B in a dosage of C mg." However, many countries do not impose such regulations because health care is a highly complex field. Imposing arbitrary external regulations may hinder physicians from fully utilizing their skills, ultimately failing to improve the quality of health care. For example, some patients with Disease A may not be able to continue on Drug B due to adverse drug reactions, while others with mild symptoms may not require Drug B. Additionally, the dosage of C mg may not be appropriate depending on factors such as body weight, kidney or liver function, or physique. These specific contents should be clarified by the guidelines of the academic society and referred to by clinicians, but they are not appropriate to be specified in law [4].

Under Japan's Medical Practitioners Act, Medical Practitioners must pass the National Medical Licensing Examination and obtain a license from the Minister of Health, Labour and Welfare. The eligibility requirements for taking the National Examination are strictly restricted to medical school graduates. Japan does not have a system for renewing medical licenses, which means that once a permit is granted, the government trusts the judgment of the licensed physician and allows them to exercise their skills as freely as possible to improve the quality of medical care [4].

However, medicine is a matter of life and death. Therefore, the government imposes minimum obligations on physicians through the Medical Practitioners Act. The act also stipulates issues related to improving the qualifications of medical practitioners, including national examinations and clinical training. The Dentists Act and the Pharmacists Act are written based on the same principles [4].

3.2 Licensing

The Minister of Health, Labour and Welfare grants a medical license to those who pass the National Medical Licensing Examination (Article 2) [1]. However, the Minister of Health, Labour and Welfare shall not grant a license to minors (Article 3, absolute disqualification reason), those who are unable to perform the duties of a physician due to physical or mental disabilities, those addicted to narcotics, cannabis, or opium, those who have been sentenced to a fine or a more severe penalty, or those who have committed criminal or fraudulent acts related to medical affairs (Article 4, relative disqualification reason) [1].

3.3 Medical School and National Medical Licensing Examination

The national examination for medical practitioners tests the requisite knowledge and skills related to medical science and public health in clinical practices that a person should possess as a medical practitioner (Article 9) [1]. A person who does not fall under any of the following items may not take the national examination for medical practitioners:

(i) a person who has graduated from a university under the School Education Act by completing a regular course in medical science (hereinafter simply referred to as "university");
(ii) a person who has passed the national preliminary examination for medical practitioners, and who has subsequently undergone practical training related to medical care or public health for one year or more; or,
(iii) a person who has graduated from a medical school in a foreign state, or who has acquired a medical practitioner's license in a foreign state, who possesses at least the same level of knowledge and skills as the persons set forth in the preceding two items and whom the Minister of Health, Labour and Welfare has approved as eligible to take the national examination for medical practitioners.

Universities or colleges with a regular course in medical science are generally called "Medical School"in Japan. However, unlike in the US, Medical School is an undergraduate school in Japan; students enter after graduating at 18 (Fig. 3.1). In Japan, every 47 prefectures has at least one medical school, and every medical school is a six-year institution. In 2025, 82 medical schools existed, including 42 national, eight public (prefectural or city governmental), 31 private, and one National Defense Medical College. Most medical schools accept 100–120 students yearly, and approximately 9300 students graduate yearly. The Ministry of Education, Culture, Sports, Science and Technology (MEXT) is responsible for policymaking related to medical education.

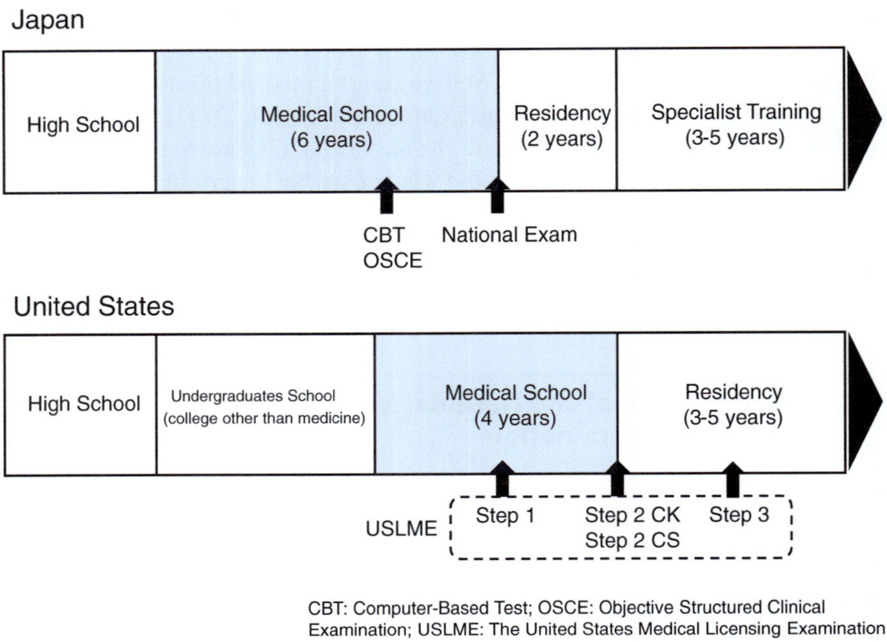

Fig. 3.1 Medical education in Japan and the US. CBT: Computer-Based Test; OSCE: Objective Structured Clinical Examination; USLME: The United States Medical Licensing Examination

A medical license is granted by registering the name of the person who has passed the National Medical Licensing Examination in the Medical Register upon their application.

Details are omitted, but rules similar to those for medical practitioners also apply to dentists and pharmacists. Medical practitioners, dentists, and pharmacists are required to report their names, addresses, and other relevant information to the Minister of Health, Labour and Welfare through the prefectural governor every two years. This notification is used for the "Survey of Medical Practitioners, Dental Practitioners, and Pharmacists" and provides critical data for public health policy-making [1–3].

3.4 Administrative Sanctions Against Physicians (Article 7)

The Minister of Health, Labour and Welfare may revoke a medical practitioner's license if the practitioner becomes subject to absolute disqualification reasons, issue a reprimand, suspend medical practice for up to three years, or revoke the license if the physician becomes subject to relative disqualification reasons or engages in conduct that undermines the dignity of a medical practitioner [1].

The Minister of Health, Labour and Welfare must hear the opinions of the Medical Ethics Council (an advisory body established within the Ministry of Health,

Labour and Welfare, composed of the President of the Japan Medical Association, the President of the Japan Dental Association, and experts in relevant fields) in advance before imposing such disciplinary actions. Since administrative disciplinary actions against physicians and dentists must be conducted fairly and impartially, it is necessary to accurately assess the facts, circumstances, and severity of the misconduct that led to the disciplinary action [1].

When determining the content of the disciplinary action, the Minister shall, as a general rule, refer to the content of judicial decisions regarding the sentencing of criminal penalties or the suspension of the execution of such penalties. Where the conduct in question violates the ethical standards expected of physicians or dentists, the Minister shall make a strict judgment, considering such factors. Additionally, the Minister of Health, Labour and Welfare may order physicians who have received administrative sanctions to undergo "reeducation training" to maintain ethical standards as a physician or acquire the knowledge and skills necessary for a medical practitioner (Article 7–2) [1].

3.5 Exclusive Right to Practice and Exclusive Use of the Title (Articles 17 and 18)

Medical care is a fundamental matter directly related to the lives and health of the public. Therefore, even if there is an agreement between the parties or an exchange of money, it is not permissible to allow medical care to be performed freely. "No person except a medical practitioner may engage in medical practice (Article 17, Exclusive right to practice)." [1]. In Japan, the Medical Practitioners Act establishes the qualifications for "Medical Practitioners," prohibits non-physicians from engaging in "medical practice," and sets forth the requirements for obtaining or losing a medical practitioner's qualifications and accompanying obligations. "Medical practice"refers to "the act of performing medical activities as a profession," where "medical activities"are defined as "activities that are dangerous unless performed with the specialized knowledge or skills of a Medical Practitioner." "As a profession" means "acts performed repeatedly and continuously toward the public or a specific group of people, regardless of accepting money." Here, "public"refers to acts performed without specifying the target, i.e., acts performed for anyone. In contrast, "specific group of people"refers to acts performed with the intent to repeat them twice or more for a substantial number of persons, such as all school students or factory employees. Even if an act is performed only once or without compensation, it is considered a "medical activity" if there is an intent to repeat it.

We must be careful that Japanese medical law strictly distinguishes between "medical care ('Iryo' in Japanese)" and "medical activities ('Igyo' in Japanese)." "Medical care"generally refers to a comprehensive concept that includes diagnosis, health guidance, and rehabilitation. Among these, acts that are dangerous if not performed by a medical practitioner, such as diagnosis, prescription, and surgery, are distinguished as "medical activities"and exclusively reserved for medical practitioners. Additionally, to ensure public trust in the title of "medical practitioner," no person except a medical practitioner may use the title of medical practitioner or a title that may be confused with this (Article 18, Exclusive Use of the Title) [1]. The strict interpretation of "medical activities"is critical in determining whether to prosecute cases where individuals practice medicine or claim to be doctors without a medical license as violations of the Medical Practitioners Act.

3.6 Dental Practitioners Act

Dentists are subject to exclusive rights to practice and use their professional title like medical practitioners. "Dental practice"refers to the act of performing medical procedures related to teeth and the oral cavity as a profession, analogous to medical practice. Prior to World War II, medical practitioners were generally authorized to practice dental medicine as part of their broader medical practice; however, when medical practitioners engaged in dental practice, they were required to undergo specific training (former National Medical Care Act, Article 8, Paragraph 2; Act, Article 34) [2].

In contrast, the current Medical Practitioners Act and Dentists Act technically exclude dental practice from the scope of medical practice, treating the two as separate professions. For example, oral surgery contains a significant portion of such practices considered medical and dental from the legal aspect; medical practitioners and dentists are permitted to engage in such practices as part of their professional duties.

3.7 Nurses and Other Co-medical Staff

While "medical practice"is defined as medical practitioners' exclusive duty, a medical practitioner alone cannot perform all medical acts in actual medical settings. Therefore, medical professionals who assist medical practitioners, namely co-medical staff, are essential.

For example, under Article 5 of the Public Health Nurse, Midwife, and Nurse Act, it defines "nurse" as a person who has acquired a nursing license from the Minister of Health, Labour and Welfare to provide medical treatment or assist in medical care for injured and ill persons or puerperal women, as a profession.

In today's highly specialized medical field, co-medical staff who assist medical practitioners in specific fields are legally defined as specialists in their respective areas. Each type of medical staff also has provisions regarding exclusive rights to perform specific duties and exclusive use of their titles [5].

In the United States, state law (not federal law) primarily governs the scope of practice and duties for licensed health professionals. For example, in Maryland, each profession's scope is defined by its licensing board under the Health Occupations Article of the Maryland Code: Maryland Board of Physicians, Maryland State Board of Dental Examiners, Maryland Board of Pharmacy, and Maryland Board of Nursing. Each board defines the scope of clinical practice (what they can legally do), licensure requirements (education, exams, continuing education), ethical and professional standards, and delegation and supervision rules (e.g., nurse practitioners, techs). Federal law plays a more significant role in areas such as Medicare/Medicaid reimbursement, controlled substances regulation (DEA licenses), and workplace laws (OSHA, HIPAA). In Japan, these rules for each profession are stipulated by national acts; therefore, the same rules apply in all prefectures.

3.8 Column: The Story of "the Separation of Medical and Dispensary Practice"

The separation of medical and dispensary practice in Europe is said to have begun when Frederick II of the Holy Roman Empire (1194–1250) feared poisoning and ordered that another person check the medications prescribed by his personal physician (Fig. 3.2). Since then, this practice has been standard in Western medicine [6].

However, in Japan, where traditional Eastern medicine, namely Kampo medicine, had a strong influence, the belief that physicians were responsible for preparing traditional Chinese medicine persisted. Even after Western medicine was

Fig. 3.2 Frederick II of the Holy Roman Empire (1194–1250)

introduced, physicians continued to prepare medications as a matter of course, resulting in the absence of pharmaceutical separation.

Before World War II, when Japan adopted various systems from Western advanced nations, it adopted the pharmaceutical separation system already in place in Germany. However, physicians were permitted to dispense medications due to a shortage of pharmacists and pharmacies, and the system did not progress in practice.

After World War II, General Headquarters/Supreme Commander for the Allied Powers (GHQ/SCAP) instructed the Japanese government to implement the American-style separation of medical and pharmaceutical services. The Ministry of Health and Welfare enacted the "Separation of Medical and Dispensary Practice Bill" in 1951. With this amendment to the law, physicians, who had been required to issue prescriptions at patients' request, are now required to issue prescriptions as a general rule with some exceptions. This situation underwent a significant change with the 1974 revision of medical fees. The prescription fee, which had been 6 points, was raised to 10 points and increased to 50 points (one point of the medical fee is equivalent to 10 JPY; see Chap. 5 for Public Health Insurance in Japan). As a result, the separation of pharmaceuticals and medical services progressed, and this year is regarded as the starting point.

References

1. Medical Practitioners' Act, Act No. 201 of July 30, 1948, Japanese Law Translation website, URL: https://www.japaneselawtranslation.go.jp/ja/laws/view/3992, ed.
2. Dental Practitioners Act, Act No. 202 of July 30, 1948, 1948, Japanese Law Translation website, URL: https://www.japaneselawtranslation.go.jp/ja/laws/view/4141, ed.
3. Pharmacists Act, Act No. 146 of August 10, 1960, Japanese Law Translation, URL: https://www.japaneselawtranslation.go.jp/ja/laws/view/2596/en, ed.
4. New Essential Forensic Medicine (6th edition, in Japanse), 2019, Ishiyaku Shuppan, ISBN978–4–263-73189-5, ed.
5. Act on Public Health Nurses, Midwives, and Nurses, Act No. 203 of July 30, 1948, Japanese Law Translation, URL: https://www.japaneselawtranslation.go.jp/ja/laws/view/3993, ed.
6. Japan Pharmaceutical Association, Iyakubungyo towa, URL: https://www.nichiyaku.or.jp/yakuzaishi/activities/division/about, ed.

4 Legal Obligations of Medical Practitioners

Jiro Ezaki, Daniel Mansour, Jun Izawa, Shoichi Maeda, and Taisuke Izumi

Chapter 3 introduced the legal definitions and concepts of medical professions. This chapter presents the legal obligations imposed on medical practitioners by the Medical Practitioners Act [1]. Dentists, pharmacists, nurses, and others also have legal obligations corresponding to their respective professional functions under their own acts [2–4].

J. Ezaki
Ministry of Health, Labour and Welfare, The Government of Japan, Chiyoda-ku, Tokyo, Japan

Department of Biology, College of Arts & Sciences, American University, Washington, DC, USA

D. Mansour
The Peter Lamy Center on Drug Therapy and Aging, Department of Practice, Science, and Health Outcomes Research, University of Maryland, School of Pharmacy, Baltimore, Maryland, USA

J. Izawa
Japan Medical Association, Bunkyo-ku, Tokyo, Japan

S. Maeda
Graduate School of Health Management, Division of Health Care Management and Public Health, Keio University, Shinjuku-ku, Tokyo, Japan

T. Izumi (✉)
Department of Biology, College of Arts & Sciences, American University, Washington, DC, USA
e-mail: izumi@american.edu

© The Author(s), under exclusive license to Springer Nature Singapore Pte Ltd. 2026

J. Ezaki et al. (eds.), *Health Policy and Law in Japan*,
https://doi.org/10.1007/978-981-95-9241-8_4

4.1 Duty to Respond to Calls and Issue Certificates (Article 19)

No medical practitioner who engages in medical practice may refuse any request for medical examination or treatment without legitimate grounds. As mentioned earlier, medical practitioners are of high public interest. They are tasked with "contributing to the improvement and promotion of public health through medical care and health guidance, thereby ensuring the healthy lives of the people." Considering the public nature of a physician's duties or the exclusivity of their professional activities, Article 17 of the Medical Practitioners Act clearly states that "a physician engaged in medical treatment shall not refuse a request for medical treatment unless there is a valid reason"(duty to respond to calls for medical assistance). "Valid reasons"are determined based on societal norms, such as when the physician themselves is ill or when they are unable to attend to other critically ill patients. Non-payment of medical fees generally does not constitute a valid reason [5].

No medical practitioner who has performed an examination or fact examination, or who was in attendance at a birth, may refuse a request for a medical certificate, certificate of postmortem examination, birth certificate, or certificate of stillbirth without legitimate grounds (Article 19-2). In many cases, these certification documents are legally and socially necessary and must not be refused without valid reasons. There is no criminal penalty for violating the obligation to respond to calls for medical assistance or issuing certificates [1].

4.2 Prohibition of Treatment Without Examination (Article 20)

No medical practitioner may provide medical treatment or issue a medical certificate or prescription without personally performing a medical examination, nor issue a birth certificate or certificate of stillbirth without personally being in attendance at the birth, nor issue a certificate of postmortem examination without personally performing the fact examination, provided, however, that this does not apply to a death certificate that has been issued where a patient undergoing medical care has died within 24 hours of the medical examination [1].

When treating a patient's disease or injury, the treatment must be appropriate based on the physician's knowledge and experience. Before deciding on treatment, the physician must examine the nature of the illness or injury. Therefore, Article 20 prohibits physicians from providing treatment without conducting an examination. Furthermore, a prescription issued to a patient enables the patient to obtain the necessary medication from a pharmacist. It thus constitutes what may be termed "indirect administration of medication," which should be regarded as one part of treatment. For this reason, it is prohibited to issue a prescription without conducting an examination [1].

Telemedicine can provide helpful information regarding the patient's physical and mental condition to replace direct face-to-face diagnosis. In that case, telemedicine does not immediately violate the prohibition on treatment without a physical examination. Diagnostic certificates, including death certificates, birth certificates, stillbirth certificates, certificates of postmortem examination, and fetal certificates of postmortem examination, are documents recognized by law as having certain legal validity. If their content is inaccurate, they may have adverse social consequences. Therefore, the Medical Practitioners Act prohibits physicians from issuing medical certificates without examining them to ensure accuracy [1].

4.3 Duty for Reporting Abnormality of a Corpse or Stillborn Baby to Police (Article 21)

When a medical practitioner finds any abnormality during the postmortem examination of a corpse or stillborn baby from a pregnancy that had entered its fourth month or beyond, the medical practitioner must notify the competent police station within 24 hours [1].

Medical practitioners should contact law enforcement authorities in such cases to facilitate criminal investigations. "The competent police station" refers to the police station with jurisdiction over the area where the autopsy was conducted. In Japan, the decision on whether to perform a forensic autopsy is generally made by the public prosecutor or police chief (see the Column: Medico-legal Death Investigation) [5].

4.4 Duty for Issuing Prescriptions (Article 22)

If a medical practitioner finds it necessary to dispense and administer medicine as a part of a patient's treatment, the medical practitioner must issue a prescription to the patient or a person caring for the patient, provided, however, that this does not apply when the patient or a person caring for the patient states that it is not necessary to issue a prescription, or in any of the following cases:

(i) when a suggestive effect is expected and issuing a prescription is likely to hinder this from being achieved;
(ii) when issuing a prescription is likely to cause anxiety to the patient about their medical care or prognosis, and may complicate the treatment of the illness;
(iii) when medicines are administered in response to changes in symptoms over short periods of time;
(iv) when the diagnosis or method of treatment has not been decided;
(v) when medicine is administered as an emergency measure necessary for treatment;
(vi) when there is no person other than the patient who requires rest, to whom the medicines can be delivered;
(vii) when stimulants are to be administered; or,
(viii) when medicine is administered aboard a vessel with no pharmacist on board.

A prescription must include the patient's name, age, drug name, dosage, administration method, date of issuance, period of use, and the name and location of the hospital or clinic, or the physician's address. It must also be signed and stamped by the medical practitioner (Fig. 4.1) [1].

Dispensing is the exclusive practice of pharmacists; however, historically, medical practitioners have often directly administered medications to their patients as part of their treatment methods in Japan (see "The Story of Separation of Medical and Dispensary Practice"). Therefore, the Pharmacists Act stipulates that persons other than pharmacists shall not dispense medications for sale or distribution, except when medical practitioners dispense drugs based on their prescriptions.

Pharmacists may not dispense drugs for sale or distribution without a prescription from a medical practitioner or other authorized person. If there are any doubts about the prescription, the pharmacist must consult the physician, dentist, or veterinarian who issued the prescription to clarify the doubts before dispensing the drugs (Pharmacists Act, Articles 23 and 24) [3].

In this way, when a physician issues a prescription to a patient, a pharmacist at a pharmacy dispenses the medication based on the prescription. Medical practitioners and pharmacists divide their responsibilities according to their respective fields of expertise to enhance the quality of national health care, a principle known as the "Separation of Medical and Dispensary Practice." (see Chap. 3, column "Separation of Medical and Dispensary Practice").

The significance of the separation of medical and dispensary practice lies in the following points: physicians can freely prescribe medications necessary for patients even if the desired medications are not readily available; patients can obtain information about the medications they are taking through the prescription; and through medication history management at a "regular pharmacy," duplicate prescriptions due to

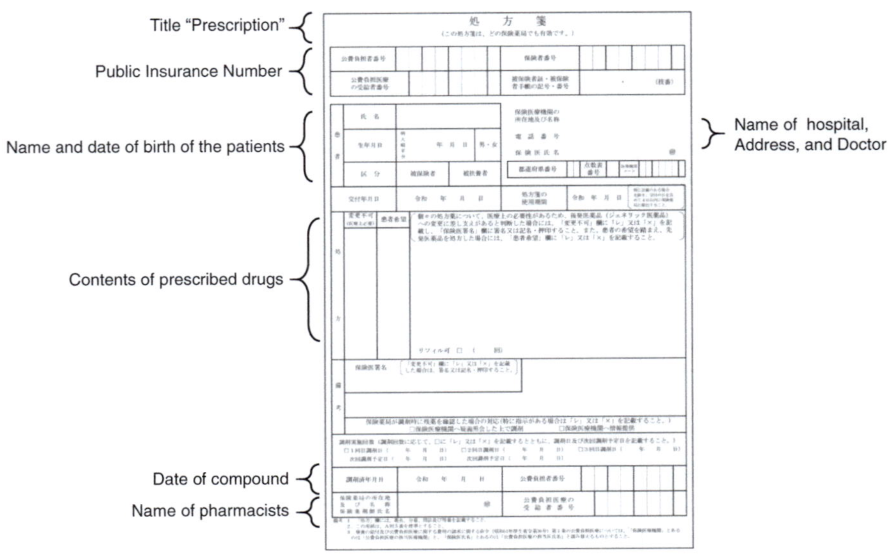

Fig. 4.1 Format and contents of a prescription

visits to multiple medical departments, confirmation of drug interactions, and other issues can be addressed, thereby improving the efficacy and safety of drug therapy.

4.5 Duty for Giving Guidance on Methods of Convalescence (Article 23)

When a medical practitioner has provided medical care, the medical practitioner must give guidance to the person in question or to their custodian on the method of treatment and other matters necessary to improve their health.

As mentioned earlier, medical practitioners are responsible for improving and promoting public health by providing medical care and conducting health guidance, thereby ensuring the healthy lives of the people (Article 1). Therefore, when treating patients, medical practitioners must not only perform surgical procedures or prescribe medications to treat illnesses or injuries but also guide convalescence and provide appropriate health guidance after recovery. Only then can they be said to have fulfilled their duties as medical practitioners [1].

4.6 Duty for Maintaining Medical Records (Article 24)

When a medical practitioner has provided medical care, the medical practitioner must enter the matters related to that medical care in a medical record without delay. Medical records as set forth in the preceding paragraph must be kept for a period of five years by the administrator of the hospital or clinic where medical care was provided by a medical practitioner works at that hospital or clinic, and by the medical practitioner themselves for medical records related to other medical care [1].

To achieve complete treatment, medical practice must be conducted consistently according to a unified plan for each patient. To this end, information related to medical treatment and therapy must be recorded and retained, and this record is legally designated as the "medical record." Medical records are also important for the administrative supervision of physicians and may occasionally serve as evidence in legal proceedings.

4.7 Medico-Legal Death Investigation

The number of decedents that police handle in Japan is approximately 200,000 per year. These include cases reported by medical practitioners as well as instances where police discover bodies, totaling 12% of all deaths (approximately 1.6 million per year).

For police-handled decedents, after postmortem examination by medical practitioners, law enforcement (prosecutor or police chief, etc.) determines the need for a forensic autopsy. Usually, forensic pathologists at the university conduct forensic autopsies. When it is clear that a person died as a result of a crime, or when death is suspected to be due to a crime, the public prosecutor, based on the Code of Criminal Procedure, may have a forensic pathologist perform an autopsy with court permission, namely judicial autopsies [5].

In cases where there is insufficient suspicion of a crime to warrant a judicial autopsy, the police chief may decide to request that a forensic pathologist conduct an "investigation autopsy." The investigation autopsy is a new category that was established in 2013 based on the Act on the Investigation of Cause of Death and on Identification of Bodies Handled by the Police [5].

In large cities with a medical examiner system (e.g., Tokyo's 23 wards (central part of Tokyo), Osaka City, Kobe City), postmortem examination and autopsies may also be performed at the medical examiner's discretion for public health reasons (medical examiner's autopsies). In these cities, the medical examiner's autopsy could be conducted on cases that were neither judicial nor investigative autopsy cases. Unlike medical examiners in the US, medical examiners in Japan basically only conduct non-criminal cases. Judicial autopsies, investigative autopsies, and medical examiner's autopsies do not require the consent of the next of kin for performing an autopsy. With consent for autopsy from the next of kin, the autopsy can be conducted when the Health Center director grants permission (see Chap. 10 for details on the Health Center). However, if a person who performs autopsies has been certified by the Ministry of Health, Labour and Welfare, the permission of the Health Center director is not required [5].

In areas without a medical examiner's system, police-affiliated medical practitioners conduct postmortem external examinations for decedents that they handle. Police-affiliated medical practitioners are not forensic pathologists but usually clinicians. The Japan Medical Association provides education to establish the professional skills required for postmortem examination among these clinicians.

In Japan, the next of kin submits a report of death to the municipality, attaching a death certificate issued by medical practitioners. Municipalities then create a Vital Statistics Survey Form, which includes information on the cause and manner of death. The form is submitted to the Minister of Health, Labour and Welfare via the public health center director and the prefectural governor. Japan's death statistics are compiled based on this data and used by the national and local public health services [5].

4.8 Japan Medical Association

This book introduces the health policies implemented by the national government, prefectures, and municipalities in Japan, as well as the laws and regulations that govern them. However, these initiatives cannot be achieved solely by the government. It is essential that various professionals, including doctors, dentists, pharmacists, nurses, public health nurses, midwives, clinical laboratory technicians, and radiological technologists, perform their duties in accordance with the principles of the system. Each profession has its professional association, which works closely with national, prefectural, and municipal governments.

For example, when the national government formulates policies related to medical care, the Japan Medical Association (JMA) participates. When prefectural governments establish policies related to nursing, members of the prefectural nursing association participate and offer their opinions. These committee members play a crucial role in enhancing the services provided to residents and patients by incorporating the perspectives of frontline staff from both professional and academic viewpoints.

The Japan Medical Association (JMA) is a notable example illustrating how professional associations contribute to the implementation of health policies. JMA is the national voice of Japanese physicians. Founded in 1916 and then re-established in its current form in 1947, the JMA's mission is to provide leadership for physicians and to promote the highest standards of medical ethics and education to protect the health of all Japanese citizens. On behalf of its members and the public in Japan, the JMA performs a wide variety of functions, such as advocating for health promotion and patient safety policies and strategies, advocating access to quality health care in local communities, and providing leadership and guidance to physicians to help them influence, manage, and adapt to changes in healthcare delivery [6].

JMA states its purpose in its Articles of Association as follows: "To promote the advancement of medical science, the development of medicine and the medical art, and the improvement of public health, in cooperation with prefectural medical associations and municipal medical associations, and thereby to contribute to the promotion of social welfare." The Articles also specify 15 different projects, including matters relating to the lifelong training of physicians, the promotion and development of community health, and the improvement of insured health care [6].

The JMA is a national academic and professional organization that consists of members from 47 prefectural medical associations. The highest decision-making body of the JMA is the House of Delegates. Delegates are elected from each Prefectural Medical Association at the rate of one delegate for every 500 members and one delegate for every fraction of that number. The regular Delegates' Meeting is held annually, and extraordinary Delegates' Meetings are convened as needed to address important matters [6].

At its establishment in 1947, the Japan Medical Association (JMA) had approximately 50,000 members; as of December 1, 2022, it had grown to 173,761 members [6]. As of December 31, 2024, membership remains relatively stable at approximately 170,000 to 175,000 physicians, representing about half of all licensed physicians in Japan which was estimated at 347,772 [7].

Prefectural Medical Associations are established in administrative areas, while Municipal Medical Associations are not necessarily limited to administrative areas. Each Prefectural Medical Association is composed of members from Municipal Medical Associations. As of the end of June 2022, there were 889 Municipal Medical Associations, including 65 University Medical Associations and 11 Prefectural Government Medical Associations. Although the Municipal Medical Associations, Prefectural Medical Associations, and the Japan Medical Association are all independent organizations, they work in close collaboration and cooperation [6].

The JMA has about 50 committees within the association, including the Committee on Bioethics, the Committee on Medical Policy, and the Council for the Promotion of Medical Science. Each committee comprises experts in medicine, economics, law, legislation, administration, and other relevant fields, and makes recommendations from a national healthcare perspective. These recommendations are reflected in administrative action through the participation of many members, as a professional medical body, in policy deliberations on national and local medical, health, and welfare issues [6].

References

1. Medical Practitioners' Act, Act No. 201 of July 30, 1948, Japanese Law Translation website, URL: https://www.japaneselawtranslation.go.jp/ja/laws/view/3992
2. Dental Practitioners Act, Act No. 202 of July 30, 1948 , 1948, Japanese Law Translation website, URL: https://www.japaneselawtranslation.go.jp/ja/laws/view/4141
3. Pharmacists Act, Act No. 146 of August 10, 1960, Japanese Law Translation, URL: https://www.japaneselawtranslation.go.jp/ja/laws/view/2596/en
4. Act on Public Health Nurses, Midwives, and Nurses, Act No. 203 of July 30, 1948, Japanese Law Translation, URL: https://www.japaneselawtranslation.go.jp/ja/laws/view/3993
5. New Essential Forensic Medicine (6th edition, in Japanese), 2019, Ishiyaku Shuppan, ISBN978-4-263-73189-5.
6. Japan Medical Association (JMA) website, overview of JMA URL: https://www.med.or.jp/english/about/overview.html
7. Ishi, Shika-ishi, Yakuzaishi toukei no gaikyou (2024) (in Japanese), Overview of the statistics of Medical Practitioners, Dental Practitioners, and Pharmacists, Ministry of Health, Labor and Welfare, Government of Japan, https://www.mhlw.go.jp/toukei/saikin/hw/ishi/24/dl/R06_kekka-1.pdf

5 Hospitals and Clinics

Jiro Ezaki, Maeve Engelbride, Shoichi Maeda, and Taisuke Izumi

To ensure appropriate medical care for the public, it is necessary to define the obligations of the "people" who provide medical care and ensure that the "places" where medical care is provided—such as hospitals and clinics—are managed, staffed, and equipped to enable the provision of high-quality and appropriate medical care. "The Medical Practitioners Act" and "the Dental Practitioners Act" govern medical professionals' status, duties, rights, and obligations; "the Medical Care Act"regulates matters related to medical facilities [1–3].

While the Medical Practitioners Act respects Medical Practitioners' freedom in the content of their medical practice to the greatest extent possible (see **Chap. 3**), the Medical Care Act imposes strict regulations on establishing and managing medical facilities, the standards for medical facilities, and matters related to advertising, medical safety, and prefectural medical care plans [4].

J. Ezaki
Ministry of Health, Labour and Welfare, The Government of Japan, Chiyoda-ku, Tokyo, Japan

Department of Biology, College of Arts & Sciences, American University, Washington, DC, USA

M. Engelbride · T. Izumi (✉)
Department of Biology, College of Arts & Sciences, American University, Washington, DC, USA
e-mail: izumi@american.edu

S. Maeda
Graduate School of Health Management, Division of Health Care Management and Public Health, Keio University, Shinjuku-ku, Tokyo, Japan

© The Author(s), under exclusive license to Springer Nature Singapore Pte Ltd. 2026

J. Ezaki et al. (eds.), *Health Policy and Law in Japan*,
https://doi.org/10.1007/978-981-95-9241-8_5

5.1 Places to Provide Medical Care: Hospitals and Clinics

Medical care must be provided as a basis for efforts to ensure and improve the health of the nation, fully respecting the wishes of the medical care recipients, and seeking efficiency and organic coordination among associated services such as welfare services, in accordance with the functions of the medical institutions (hospitals, clinics, long-term care health facilities, integrated facilities for medical and long-term care, dispensing pharmacies, and other facilities that provide medical care) and in the homes of medical care recipients, etc. (Article 1–2, Medical Care Act; the same shall apply hereinafter) [4].

In addition to the homes of medical care recipients, "etc." includes facilities specified under the Elderly Welfare Act, such as nursing homes, special nursing homes, low-cost nursing homes, paid nursing homes, and other places where persons receiving medical care can lead a convalescent lifestyle.

Under the Medical Care Act, the places where medical practitioners or dental practitioners practice medicine or dentistry for the public or a specific number of people are limited to "hospitals" and "clinics." The distinction between hospitals and clinics is as follows: hospitals must have 20 or more beds, while clinics must have no beds or 19 beds or fewer. In English, the terms "hospital" and "clinic" are not always distinguished; however, in Japan, they are strictly distinguished based on the number of beds [4].

A "hospital" is a facility capable of providing truly scientific and appropriate medical treatment to patients, with sufficiently well-developed structural and equipment standards. Certain hospitals with specific functions are designated as "advanced treatment hospitals," "regional medical care support hospitals," or "core hospitals for clinical research." Different requirements, including staffing standards, structural and equipment standards, and the responsibilities of administrators, are established for these types of hospitals [4].

5.2 Midwifery Center

A "midwifery center" refers to a facility where midwives perform their duties for the public or a specific group of people (Article 2). The services provided at a midwifery center are the duties of a midwife, namely assisting with childbirth and providing health guidance to pregnant women, new mothers, and newborns. A midwifery center differs from a clinic because it does not perform medical practice. Normal childbirth is not considered to be medical care. Therefore, no doctors are on staff, and medical practice cannot be performed at a midwifery center. As a result, if medical practice, such as examinations, is necessary, they will be performed by a medical practitioner or medical institution (hospital or clinic) contracted by the midwifery clinic as appropriate [4].

5.3 Medical Inspections

To ensure the appropriateness of facilities and equipment at hospitals, clinics, and maternity facilities, medical care inspectors, mostly health center staff, are appointed to conduct regular and irregular on-site inspections and provide guidance to medical institutions (Article 25, 26) (medical inspection) [4].

Furthermore, to ensure the effectiveness of the standards outlined in the Medical Care Act, provisions are made for restrictions on the use of facilities, inspections prior to use, orders to change managers, revocation of licenses to operate hospitals, and orders to close facilities (Articles 27 to 29) [4].

5.4 Restrictions on Advertising

No party, when placing any advertisement or making any other representation as a means of inducing medical care recipients with regard to medical practice or dental practice, or hospitals or clinics, is to run false advertisements, regardless of whether it is made in writing or by any other means.

The contents and methods of the advertisement must conform to the following standards so as not to hinder the appropriate selection of medical care by medical care recipients:

(i) No advertisement is to be placed to the effect that the hospital or clinic is superior to other hospitals or clinics.
(ii) No exaggerated advertisement is to be placed.
(iii) No advertisement that undermines public order or corrupt good morals is to be placed.
(iv) Other standards specified by an Order of the Ministry of Health, Labour and Welfare as necessary for appropriate selection of medical care. (Article 6–5 of the Act)

It would be highly convenient for citizens to know which physicians practice medicine, the medical services they offer, and the primary diseases they treat if they contract an illness. From this perspective, it is beneficial for the public to allow medical practices to advertise their services in as much detail as possible if it is truthful. However, medical services involve human life and physical well-being, and improper advertising may cause recipients to receive inappropriate services, resulting in significantly more severe harm than in other fields. Furthermore, because medical services are highly specialized, recipients of advertising find it extremely difficult to assess the quality of services based solely on the wording of the advertisement (Fig. 5.1) [4].

For these reasons, medical advertising is generally prohibited in order to protect patients and other users. Some basic information is allowed to be advertised, such as the medical or dental practitioner's name or specialty, the name, phone number, or location of the hospital or clinic, and the name of the hospital or

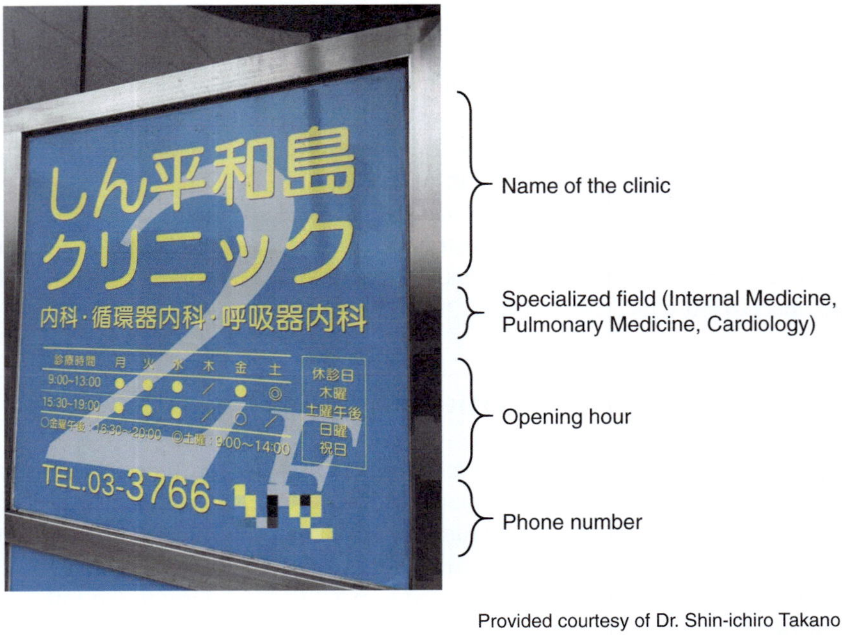

Fig. 5.1 Example of a Medical Institution Advertisement. (Provided courtesy of Dr. Shin-ichiro Takano)

clinic's administrator. The Ministry of Health, Labour and Welfare has established the "Medical Advertising Guidelines," which provide specific examples and clarify what types of medical advertising are prohibited. They have also established rules regarding the names of medical departments that can be advertised. Hospitals and clinics must comply with these rules when advertising their medical departments [4].

In Japan, once medical practitioners have obtained a medical license, they can freely choose their specialties and advertise them externally, regardless of their specialty or years of experience. In other words, medical practitioners can personally decide which medical services to provide and communicate this information to patients. However, for anesthesiology, only physicians who have obtained permission from the Minister of Health, Labour and Welfare (anesthesiology-designated physicians) may designate themselves as such. To become an anesthesiology-designated physician, specific training and experience are required [4].

5.5 Postwar History

World War II caused significant damage to people's lives and economies, and the medical field was no exception. Many medical facilities were destroyed or closed due to war damage, and combined with shortages of medical personnel, food, medicines, and sanitary materials; the post-war situation of medical facilities in Japan was dire [5].

In 1945, the Ministry of Health and Welfare (at the time) received former Japanese military hospitals from the occupying forces and decided to open them to the general public as national hospitals and sanatoriums. This had significant implications in addressing the shortage of medical facilities [5].

The Medical Care Act (enacted in 1948) established facility standards for public hospitals operated by prefectures and municipalities to ensure the quality of medical care. It also provided provisions allowing the national government to subsidize the establishment of such hospitals. In 1951, the scope of national government subsidies was expanded to include other public medical institutions such as the Japanese Red Cross Society, National Welfare Federation of Agricultural Cooperatives, and Social Welfare Organization Saiseikai Imperial Gift Foundation, Inc. These public hospitals continue to provide medical care in many areas today. Hospital development progressed year by year, helped by the economy's recovery and the growing momentum for hospital construction among local governments. The number of municipally operated clinics rose significantly due to increased national subsidies for their development. Other public hospitals also gradually recovered. However, while the development of these hospitals significantly improved the standard of medical care, each institution carried out facility expansion based on its own judgment, making it difficult to say that this was carried out under a comprehensive plan [5].

5.6 Medical Corporation

Due to the severe economic conditions following the war, it was difficult for individual physicians to establish private hospitals. Under the Medical Care Act, establishing for-profit medical institutions is prohibited, and medical corporations are prohibited from distributing surplus funds. While for-profit companies could incorporate as legal entities in other sectors, health care is a non-profit sector without such a means of incorporation [5].

To facilitate funding, the Medical Corporation System was established in 1950 under the Medical Care Act, allowing hospitals to obtain non-profit legal status. As a result, stable medical facility management became possible, and the establishment of private hospitals progressed, particularly in urban areas. Specifically, over the 10 years from 1955, the number of beds in private hospitals nearly doubled from 198,096 to 424,224 [5].

5.7 Prefectural Medical Care Plan

In 1956, the Social Security System Council pointed out the importance of developing a network of medical institutions, including private ones, while raising concerns about the excessive proliferation of public medical institutions. As a result, in 1962, the Medical Care Act was revised to impose an obligation for national and local governments to develop hospitals and clinics in areas with a shortage of medical institutions. At the same time, they introduced a system of bed restrictions for public hospitals. In 1986, the Medical Care Act stipulated that prefectures should formulate medical care plans, set the necessary hospital beds for each "secondary medical area," and prevent the establishment of new hospitals and beds exceeding the necessary number. A "medical area"is a regional unit established by prefectural governments to ensure a system for providing medical care tailored to the region's conditions. Medical areas are divided into primary medical care areas (basically municipalities), which provide health care closely linked to daily life; secondary medical care areas (multiple municipalities), which provide general health care ranging from promoting healthy lifestyles and disease prevention to inpatient treatment; and tertiary medical care areas (basically prefectural units), which provide specialized medical care requiring advanced technology [4].

It was decided to promote the development of public and private medical facilities by creating medical care plans tailored to regional circumstances. Prefectural governors were authorized to set bed quotas for secondary medical areas, alongside the existing quotas for public hospitals, and to make recommendations for the establishment or expansion of hospitals in regions with excess beds. These recommendations were to occur after consulting with the Prefectural Medical Care Council, who set the necessary number of beds per medical care area.

The Prefectural Medical Plan includes provisions on medical collaboration systems for five major diseases—cancer, stroke, acute myocardial infarction, diabetes, and mental disorders—and five key medical services: emergency medical care, disaster medical care, remote area medical care, perinatal medical care, and pediatric medical care as well as responses to emerging infectious diseases. This plan also covers home-based medical care; medical personnel; ensuring medical safety; goals for the development of medical facilities; target number of beds; and regional medical plans (Article 30-4).

5.8 Functional Differentiation of Hospitals and Beds

In 1992, the "advanced treatment hospital"category was established to promote the functional differentiation of medical facilities. "Advanced treatment hospitals"provide advanced medical services, develop advanced medical technologies, and accept patients referred from other hospitals or clinics. They are primarily designated as university hospitals [4, 5].

In 1997, a system was established to designate certain hospitals as "Regional Medical Support Hospitals"to support primary care physicians and dentists. These hospitals provide medical care to referred patients, share or open their facilities and equipment, conduct emergency medical care, and train local medical personnel [4, 5].

In 2000, beds other than psychiatric, infectious disease, and tuberculosis beds were classified into "long-term care beds," designed to provide a convalescent environment for patients in the chronic phase, or "general beds," which have a higher ratio of doctors and nurses.

5.9 The Provision of Information to Patients

In providing medical care, a physician, dentist, pharmacist, nurse, or other medical care professional must give proper explanations and endeavor to foster understanding in medical care recipients (Article 1–4(2)) [4].

The Medical Care Act explicitly stated the principles of medical care, stipulating that medical care must be based on respect for life and the dignity of the individual, founded on a relationship of trust between medical providers and recipients, and of high quality and appropriate to the recipient's physical and mental condition. Subsequently, amendments to the Medical Care Act were successfully enacted to promote high-quality and proper medical care based on such trust, including measures to enhance patient information provision.

Provisions were established in 1997 requiring healthcare providers to provide patients with reasonable explanations and ensure their understanding when delivering medical care. Thus, an obligation to strive for informed consent, including adequate explanation and consent from patients, was established.

In 2000, regulations on medical institution advertising were relaxed in response to the growing need for patients to access information about their medical care and to facilitate patients' choice of medical institutions. Specifically, the following items were added as matters that could be included in advertising: "Information regarding medical records, etc., may be provided," and "Results of medical care evaluations conducted by the Japan Medical Care Evaluation Organization" [4].

5.10 Medical Accident Investigation

The Medical Accident Investigation System was established in 2015 to ensure medical safety and prevent the recurrence of medical accidents (Article 6–10). The hospital administrator must report any medical accidents to the Medical Accident Investigation and Support Center. Medical accidents are defined as deaths or stillbirths caused by or suspected to be caused by medical care provided that the administrator did not anticipate (Fig. 5.2) [4].

The hospital administrator must immediately investigate the medical accident to clarify the cause. Upon completion of the investigation, the administrator must explain the results to the bereaved family and report them to the Medical Accident Investigation and Support Center. Additionally, the Medical Accident Investigation and Support Center must investigate and report the results to the healthcare facility administrator and the bereaved family when requested by the healthcare facility administrator or the bereaved family regarding cases reported to the center as "medical accidents" [4].

Fig. 5.2 Scheme of the Medical Accident Investigation

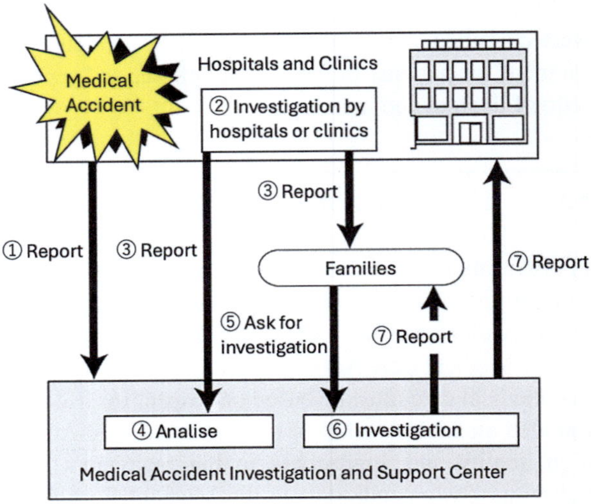

References

1. Medical Practitioners' Act, Act No. 201 of July 30, 1948, Japanese Law Translation website, URL: https://www.japaneselawtranslation.go.jp/ja/laws/view/3992
2. Dental Practitioners Act, Act No. 202 of July 30, 1948 , 1948, Japanese Law Translation website, URL: https://www.japaneselawtranslation.go.jp/ja/laws/view/4141
3. Pharmacists Act, Act No. 146 of August 10, 1960, Japanese Law Translation, URL: https://www.japaneselawtranslation.go.jp/ja/laws/view/2596/en
4. Medical Care Act, Act No. 205 of July 30, 1948, Japanese Law Translation, URL: https://www.japaneselawtranslation.go.jp/ja/laws/view/4006
5. Kousei-roudou hakusho (2007) (in Japanese) (Ministry of Health, Labour and Welfare's White Paper), p. 4–27, ISBN-13. 978-4324083031., data available from URL: https://www.mhlw.go.jp/wp/hakusyo/kousei/07/dl/0101.pdf

Clinical Trials and Drug Regulations

6

Jiro Ezaki, Shoichi Maeda, and Taisuke Izumi

Rules are necessary in medical care, not only for medical professions, as defined by the Medical Practitioners Act, etc. (see Chaps. 3 and 4), and hospitals and clinics, as defined by the Medical Care Act (see Chap. 5), but also for pharmacies, drugstores, and the efficacy and safety of drugs and medical devices [1, 2].

The Pharmaceutical Affairs Act, enacted in 1960, has established regulations to ensure the quality, effectiveness, and safety of pharmaceuticals, quasi-pharmaceuticals, cosmetics, and medical devices. Major amendments were made to the Act on Securing Quality, Efficacy, and Safety of Products, Including Pharmaceuticals and Medical Devices Act (PMD Act) in 2013 [3]. This amendment established independent chapters for medical devices, distinct from those for pharmaceuticals. It also covers the manufacture and sale of regenerative medical products which are related to gene or cell therapy for disease treatment or prevention under the same act [3].

This chapter explains the Pharmaceutical and Medical Device Act using pharmaceuticals as an example. However, medical devices and regenerative medical products are subject to similar legal structures. This chapter overviews the PMD Act, the Clinical Trials Act (2018), and the Ethical Guidelines for Medical and Biological Research Involving Human Subjects (2021) [3–5].

J. Ezaki
Ministry of Health, Labour and Welfare, The Government of Japan, Chiyoda-ku, Tokyo, Japan

Department of Biology, College of Arts & Sciences, American University, Washington, DC, USA

S. Maeda
Graduate School of Health Management, Division of Health Care Management and Public Health, Keio University, Shinjuku-ku, Tokyo, Japan

T. Izumi (✉)
Department of Biology, College of Arts & Sciences, American University, Washington, DC, USA
e-mail: izumi@american.edu

© The Author(s), under exclusive license to Springer Nature Singapore Pte Ltd. 2026

J. Ezaki et al. (eds.), *Health Policy and Law in Japan*,
https://doi.org/10.1007/978-981-95-9241-8_6

6.1 Categories of Pharmaceuticals

Under the PMD Act, pharmaceuticals are broadly categorized into "prescription drugs (Rx)" and "over-the-counter (OTC) pharmaceuticals" [3]. Prescription drugs are pharmaceuticals supplied with their prescriptions or instructions by physicians or dentists, whereas OTC drugs are pharmaceuticals other than those classified as prescription drugs. OTC drugs are classified into three categories—Category 1, Category 2, and Category 3—in order by risk levels (1 > 2 > 3) [3]. A new category, "Pharmaceuticals Requiring Guidance," has been established to bridge the distinction between prescription and OTC drugs. Prescription drugs are sold only through face-to-face sales, while over-the-counter drugs can be sold online or through other means [3].

6.2 Criteria of License for Selling Pharmaceuticals

The PMD Act establishes the requirements and licensing procedures for four types of drug sales businesses: pharmacies, retail drug stores, distribution pharmacies, and wholesale pharmacies [3].

A "pharmacy" is a facility that dispenses medicine, provides information related to the pharmaceutical, and gives guidance for medication. "Dispensing" refers to compiling pharmaceuticals according to a prescription and providing medication guidance to patients as part of medical care. Therefore, pharmacies are classified as "medical care facilities" alongside hospitals and clinics under the Medical Care Act (see Chap. 5). For this reason, above, pharmacies must always have a pharmacist on duty and have a dispensing room. Additionally, pharmacies are exclusively authorized to handle prescription drugs based on prescriptions. Of course, pharmacies can sell OTC drugs as well as prescription drugs [3].

A "store-based distribution," generally referred to as drug stores, is permitted to sell prescribed drugs by pharmacists, any OTC drugs regardless of categories by pharmacists, and Categories 2 and 3 by a registered salesclerk. In store-based distribution, they cannot perform dispensing. In addition, for OTC drugs that meet specific criteria, such as being less prone to changes over time, a system where boxes containing medications are pre-placed in households, and the cost of used medications is settled during the next visit, is referred to as "Household Distribution" [3]. The household distribution in Toyama has a history spanning over 300 years. "Sen'yō Kōri" (literally, "use first, pay later") is a sales method mainly used for over-the-counter medicines in Toyama. Customers use the product first and pay for it later based on how much they have used. This phrase embodies the spirit of Toyama medicine sellers, who prioritize the health of their customers and put profits second.

Furthermore, selling or dispensing pharmaceuticals exclusively to pharmaceutical operators, wholesalers, pharmaceutical manufacturers, medical institutions, and other entities is defined as "Wholesale Distribution" [3].

6.3 Licensing System and Pharmaceutical Approval

Pharmaceutical companies need a "Marketing License," a "License for Manufacturing," and a "Marketing Approval for Pharmaceuticals" from the Ministry of Health, Labour and Welfare to manufacture and sell pharmaceuticals. In regard to the Marketing License, manufacturers and distributors must comply with quality control standards (GQP) and post-marketing safety management standards (GVP), and appoint a general manager who is in charge of manufacturing and sales [3]. For the License for Manufacturing, each manufacturing facility is granted a manufacturing license based on its type of manufacturing, and the facility must meet the structural and equipment standards specified in the Pharmacy Structural and Equipment Regulations [3].

The Minister of Health, Labour and Welfare grants approval for the manufacturing and distribution of pharmaceuticals upon application by the entity seeking such approval. The approval process involves reviewing each product item based on its name, ingredients, dosage, administration method, efficacy, effects, side effects, etc., by the current academic standards in medicine, pharmacy, and related fields. Additionally, for certain pharmaceuticals, compliance with the standards for manufacturing and quality control methods (GMP) is a prerequisite for approval [3].

6.4 PMDA Examination on Pharmaceuticals

The review for marketing approval of new pharmaceuticals requires cautious consideration of their efficacy and safety, so extensive submission of basic data, clinical data, and quality-related data is mandatory [3].

After the Pharmaceutical and Medical Devices Agency (PMDA) conducts the necessary evaluations, the matter is referred to the Pharmaceutical Affairs Council, an advisory body to the Minister of Health, Labour and Welfare, for review by experts in relevant fields. The Minister of Health, Labour and Welfare decides whether to approve or reject the new pharmaceuticals following the council's recommendations (Fig 6.1) [3].

Clinical studies are conducted to confirm the effects and side effects of drugs when they are used and to collect data for submission as part of the application for approval of a pharmaceutical product. Generally, studies evaluating the efficacy and safety of pharmaceutical products are called "clinical studies." However, among clinical studies, studies conducted to collect data for obtaining national approval using "drug candidates" are specifically referred to as "clinical trials" to distinguish them from other types of clinical studies and must comply with the GCP Ordinance (Ordinance on Standards for Conducting Clinical Trials on Pharmaceuticals).

Under the PMD Act, clinical trial sponsors (pharmaceutical companies, etc.) and medical practitioners (those who conduct clinical trials themselves) are required to submit a clinical trial plan to the Minister of Health, Labour and Welfare and report any adverse effects, malfunctions, etc., observed during the clinical trial.

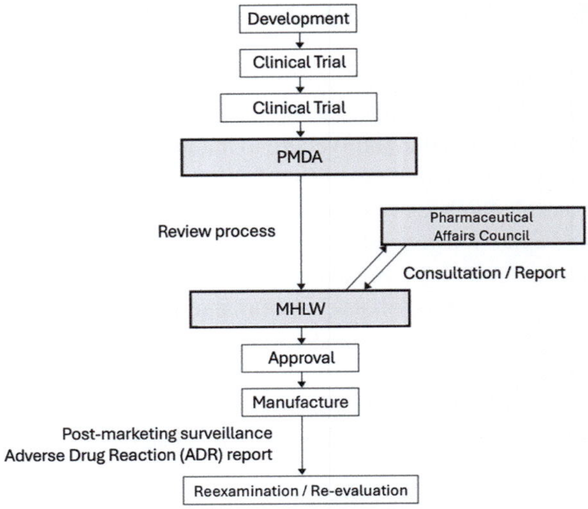

Fig. 6.1 The review process for marketing approval of new pharmaceuticals

Pharmaceutical companies typically conduct clinical trials (company-sponsored clinical trials) [3].

The responsible medical practitioners at a medical institution that a pharmaceutical company or other entity has commissioned to conduct a clinical trial organize a team with other medical practitioners conducting the trial (sub-investigators), clinical research coordinators (CRCs), and clinical trial administrators and collaborate with departments such as nursing, pharmacy, and laboratory services to conduct the trial appropriately. On the other hand, there are also clinical trials conducted primarily by physician or dentist, known as physician-initiated clinical trials [3].

For both types of clinical trials, the sponsor (company or physician) must submit a clinical trial plan notification (CTN) to the PMDA prior to initiation of the clinical trial. For new active ingredients, new routes of administration, new combination drugs, etc., the PMDA conducts a 30-day investigation from the submission date to confirm the safety of the clinical trial. If no issues are raised during the 30-day investigation, the clinical trial can be initiated [3].

6.5 Approval Process for the Orphan Drugs

Developing new drugs requires substantial funding, a lengthy timeframe, and significant development risks. It is incredibly challenging to rely on the voluntary efforts of private companies to research and develop drugs, medical devices, and regenerative medicine products that have limited patient numbers, low market potential, and other similar characteristics. Therefore, the national government must actively promote research and development to reduce the burden on companies and promote the development of drugs, medical devices, and regenerative medicine products that are medically necessary but have not progressed due to the small number of patients so that orphan drugs, etc. can be made available in medical settings as soon as possible [3].

The Orphan Drug Development Promotion System was established in 2013 to support the development of orphan drugs. Additionally, based on the Pharmaceutical and Medical Devices Act, the implementation of orphan regenerative medicine products began in 2014. The preferential measures applicable when designated include (1) grant disbursement, (2) guidance and advice, (3) tax credit for research and development expenses, (4) priority review, and (5) extension of the re-review period [3].

6.6 Designated Marketing Authorization Holder (DMAH)

When a foreign manufacturer sells a product in Japan, even if it does not have a place of business in Japan, it can become an applicant for approval and hold a Japanese approval on its own (Marketing Approval for Pharmaceuticals in Foreign Countries, Article 19-2). In such cases, a domestic marketing authorization holder must be appointed as the marketing authorization holder (Designated Marketing Authorization Holder (DMAH)). DMAHs are obligated to conduct manufacturing and quality control in Japan by the QMS Ordinance, safety control operations by the GVP Ordinance, and implementation and compliance with operations related to the product. A similar system is applied to medical devices (Article 23-2-17(1)) [3].

6.7 Specified Clinical Trials Under the Clinical Trials Act

Clinical trials must comply with the PMD Act and Good Clinical Practice (GCP) regulations. However, clinical studies that do not fall under this category are not subject to legal restrictions and have been conducted based on the governmental guidelines called "Ethical Guidelines for Clinical Research." The guideline establishes researchers' responsibilities regarding informed consent from participants, protection of personal information, and matters related to review by an ethics review committee. However, even if the guidelines were violated, the government could only issue non-binding administrative guidance [4].

In Japan, in 2012, cases of misconduct that undermined the reliability of clinical studies, such as data tampering by pharmaceutical companies, inadequate consent from participants, and the leakage of personal information, occurred. The causes of the misconduct were identified as insufficient regulations compared to those in Western countries and a lack of sufficient supervisory authority by administrative agencies. Therefore, the Clinical Trials Act was enacted in 2018 to establish procedures for conducting clinical studies and a system for disclosing information on funding for clinical studies [4].

At this point, we must be careful that the term "clinical trials" has two different meanings in the Japanese medical law: (a) one is a clinical trial under the PMD Act, which means clinical trials for approval applications ("Chiken" in Japanese), and (b) the other is clinical trials under the Clinical Trials Act, which does not include the clinical trials under the PMD Act ("Rinshou-kenkyu" in Japanese, which corresponds to clinical studies in the US). Since people translate two different Japanese terms ("Chiken" and "Rinshou-kenkyu") into the identical English term "clinical trial," everyone must be aware of the context in which they are discussing clinical trials [6].

The Clinical Trials Act defines "specified clinical trials" as those involving unapproved or off-label drugs, or clinical trials on drugs funded by pharmaceutical companies [4]. For "specified clinical trials," the Act requires trial conductors to establish implementation plans, trial protocols, conflict of interest management plans, and other documents, and comply with implementation standards for monitoring and auditing, conflict of interest management, obtaining informed consent, protecting personal information, and preserving records [4]. Additionally, the Minister of Health, Labour and Welfare may issue improvement orders for a clinical study that violates implementation standards. In case investigators don't follow the orders, the Minister of Health, Labour and Welfare may force them to suspend the trials [4].

6.8 Ethics for Life Science and Medical Research

In 1964, the World Medical Association (WMA) adopted the Declaration of Helsinki as ethical principles for medical research involving human subjects. The Declaration has been revised in response to advances and diversification in medical research. It stipulates provisions such as comparing risks and benefits to participants, preparing research protocols, approval by an ethics committee, and informed consent (where participants receive adequate prior explanation from researchers, understand the significance, purpose, and methods of the research, and provide consent based on their free will).

The Ministry of Health, Labour and Welfare has formulated ethical guidelines for various types of research in collaboration with relevant ministries and agencies. These guidelines consider the ethical norms outlined in the Declaration of Helsinki and discussions on the protection of personal information in Japan. Additionally, in response to the diversification of research subjects and methods and advancements in life sciences, medicine, and medical technology, the scope and methods of regulation are subject to ongoing review.

In 2021, the existing ethical guidelines were consolidated into the "Ethical Guidelines for Life Science and Medical Research Involving Human Subjects." Even research that does not fall under clinical trials under either the PMD Act or the Clinical Research Act must comply with ethical guidelines when it involves human subjects in life science and medical research [5].

References

1. Medical Practitioners' Act, Act No. 201 of July 30, 1948, Japanese Law Translation website. https://www.japaneselawtranslation.go.jp/ja/laws/view/3992.
2. Medical Care Act, Act No. 205 of July 30, 1948, Japanese Law Translation. https://www.japaneselawtranslation.go.jp/ja/laws/view/4006.
3. Act on Securing Quality, Efficacy and Safety of Products Including Pharmaceuticals and Medical Devices (PMD act), Act No. 145 of August 10, 1960, Japanese Law Translation. https://www.japaneselawtranslation.go.jp/ja/laws/view/3213.
4. Clinical Trials Act, Act No. 16 of April 14, 2017. https://h-crisis.niph.go.jp/wp-content/uploads/2019/03/20190319105245_file_06-Seisakujouhou-10800000-Iseikyoku_0000213334.pdf.
5. Ethical Guidelines for Medical and Biological Research Involving Human Subjects (2021) Ministry of Education, Culture, Sports, Science and Technology (MEXT); Ministry of Health, Labour and Welfare (MHLW); Ministry of Economy, Trade and Industry (METI). https://www.mext.go.jp/content/20250325-mxt_life-000035486-01.pdf.
6. Kuwahara H. Overview of the Clinical Trial Act. J Jpn Soc Clin Anesth. 2019;39(4):416–22. (in Japanese).

7 Health Insurance

Jiro Ezaki, Hasset Tibebe, Shoichi Maeda, and Taisuke Izumi

After World War II, under the guidance of GHQ, the Medical Practitioners Act and the Medical Care Act were enacted, laying the foundation for modern healthcare systems (see Chaps. 3, 4, 5, and 6). However, many citizens could not afford medical care due to poverty. This chapter focuses on the universal healthcare system established in Japan, which was implemented in 1961 [1].

The origin of Japan's medical insurance system dates to the Health Insurance Act of 1922. The Act was established as a labor policy to address growing concerns about the livelihoods of workers amid rapid workforce expansion and to enhance labor productivity. Subsequently, the former National Health Insurance Act, primarily aimed at assisting farmers, was enacted in 1938. At this time, not all citizens were covered by public health insurance. However, this public health insurance system played a critical role as the basis of the universal health insurance system, which was established after World War II [1].

As of March 31, 1956, approximately 28.71 million people (about 32% of the total population at that time) remained uninsured under the medical insurance system. Uninsured individuals primarily consisted of workers in small and medium-sized enterprises, as well as those in the agricultural, forestry, and fisheries

J. Ezaki
Ministry of Health, Labour and Welfare, The Government of Japan, Chiyoda-ku, Tokyo, Japan

Department of Biology, College of Arts & Sciences, American University, Washington, DC, USA

H. Tibebe · T. Izumi (✉)
Department of Biology, College of Arts & Sciences, American University, Washington, DC, USA
e-mail: izumi@american.edu

S. Maeda
Graduate School of Health Management, Division of Health Care Management and Public Health, Keio University, Shinjuku-ku, Tokyo, Japan

© The Author(s), under exclusive license to Springer Nature Singapore Pte Ltd. 2026

J. Ezaki et al. (eds.), *Health Policy and Law in Japan*,
https://doi.org/10.1007/978-981-95-9241-8_7

sectors, and self-employed individuals. "The dual structure," namely the medical disparity between large-enterprise workers and those in small- and medium-sized enterprises, as well as between residents of municipalities with National Health Insurance and those without, was a significant issue. To address these issues, the government proposed the "New National Health Insurance Act" to strengthen the National Health Insurance system and establish the foundation for universal health insurance. The National Diet passed the bill in December 1958. The law came into effect in January 1959, and as initially planned, the universal health insurance system was fully implemented by April 1961 [1].

The National Health Insurance Act of 1961 established a medical care system in Japan that ensures that everyone can receive medical care with peace of mind, marking a significant step forward toward achieving the world's highest average life expectancy and high standards of health and medical care [1].

7.1 National Health Insurance Act

The National Health Insurance Act requires municipalities to operate national health insurance programs. It involves everyone who is not covered by other public health insurance to enroll in the national health insurance program of their municipality [2]. In other words, everyone is enrolled in some form of public health insurance and pays premiums to the insurer, namely Universal Health Coverage (UHC). Currently, there are approximately 3000 insurers, including municipalities, union health insurance plans, and mutual aid associations. As a result, insured citizens can receive medical treatment at any hospital or clinic, pay a portion of the costs out of pocket, and receive medical benefits (including free access).

The structure of insured medical care under Japan's health insurance system is as follows (Fig. 7.1):

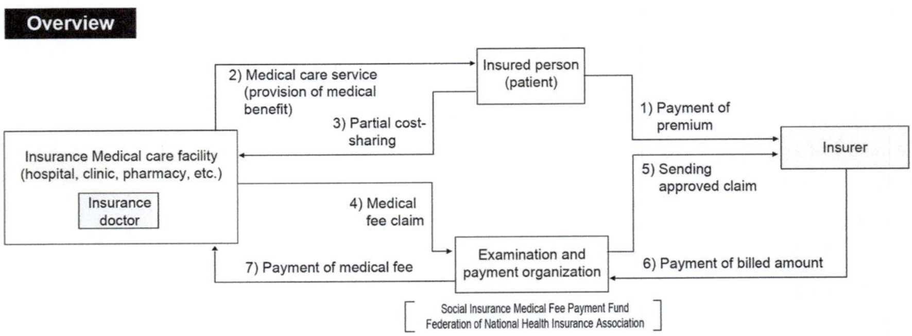

Medical fees are classified into three types: medical, dental, and dispensing fees.
The medical fee is calculated by adding stipulated numbers of points for the individual medical activities provided (so-called "fee-for-service system"). The unit price for one point is ¥10. For a typhlitis hospitalization case, for example, the first visit fee, the hospitalization fee multiplied by the length of stay (days), the typhlitis surgery fee, the test fee and the drug fee are added to one another and insurance medical care facility will receive the total amount less the patient's co-payment from the examination and payment organization.

Fig. 7.1 Conceptual chart of insured medical treatment

① Insured individuals pay premiums to the insurance provider. The health insurance provider registers the insured individual's information in an online eligibility verification network.
② When an insured individual becomes ill or injured, they receive medical services (medical care benefits) at an insured medical facility (hospital, clinic, etc.).
③ The insured individual pays a portion of the costs (copayment) when receiving medical services.
④ The insured medical facility submits a claim for medical fees to the review and payment agency.
⑤ The review and payment agency reviews the claims submitted by medical institutions and then claims the amounts from the health insurance provider.
⑥ The health insurance provider pays the claimed amounts to the review and payment agency.
⑦ The review and payment agency pays the medical fees to the medical institutions.

7.2 Ordinance on Health Insurance

Only insurance medical care facilities that have applied for designation under the Health Insurance Act (Article 65 of the Health Insurance Act) and have been designated by the Minister of Health, Labour and Welfare may provide medical care covered by health insurance (Article 63 of the Health Insurance Act) [3].

Furthermore, only medical and dental practitioners or pharmacists registered under the Health Insurance Act are permitted to provide medical care covered by health insurance at these medical institutions or pharmacies (Articles 64 and 71 of the Health Insurance Act). In other words, medical and dental practitioners, as well as pharmacists, do not automatically acquire the qualifications of insured physicians or insured pharmacists merely by passing the national examination and obtaining a license. Through this registration, physicians, dentists, and pharmacists are obligated to provide medical care under the health insurance system by the Ministry of Health, Labour and Welfare's ordinance (Medical Care Provision Regulations).

As mentioned earlier, when patients visit an insured medical care facility, they pay a portion of the costs, namely a copayment, at the hospital or clinic's front desk, and the insurance provider covers the remaining costs through the review and payment agency. This system is stipulated in the Health Insurance Act and other relevant laws and is implemented by insurance-covered medical institutions that voluntarily participate in these provisions. From this perspective, insurance-covered medical treatment is also referred to as "contractual medical treatment based on a public legal contract between the insurance provider and the insurance-covered medical institution."

The following outlines the major rules in insurance-covered medical treatment in an insured medical care facility. The provisions of the Medical Practitioners Act, the Dental Practitioners Act, the Pharmacists Act, the Medical Care Act, and the Pharmaceutical Affairs and Medical Devices Act, among others, remain in effect regardless of whether the medical treatment is insurance-covered. On the other hand, the Medical Care Regulations are provisions that must be complied with in addition to the laws mentioned above when providing insurance-covered medical treatment [4].

First, insured medical care facilities must not induce patients to receive treatment at their facilities by providing economic benefits that could undermine the sound operation of the health insurance system, such as offering discounts on goods related to revenue-generating activities based on the amount of fees paid by patients for treatment.

Additionally, acts such as providing monetary or other benefits to business operators or their employees in exchange for patient referrals are also prohibited for the same reason (Medical Care Regulations Article 2-4-2 Prohibition of Inducement Through the Provision of Economic Benefits). Health checkups should not be considered insured medical care.

Furthermore, examinations, medication, injections, surgeries, or procedures must be performed after fully considering the medical necessity, and excessive treatment (overdiagnosis) is prohibited (Ordinance on Health Insurance Article 20).

Additionally, the implementation of "special therapies or new therapies, etc." for which medical evaluation is not yet sufficiently established, the use of "drugs other than those specified by the Minister of Health, Labour and Welfare," and the conduct of examinations for "research purposes" are not recognized under health insurance coverage (Ordinance on Health Insurance Article 20. Articles 18 to 20).

The terms "drugs specified by the Minister of Health, Labour and Welfare" refer to drugs listed in the Drug Price Standards (as described below) or drugs recognized as "Medical Expenses Combined with Treatment Outside Insurance Coverage."

7.3 Medical Expenses Combined with Treatment Outside Insurance Coverage

Medical Expenses Combined with Treatment Outside Insurance Coverage refers to medical care or medications not covered by insurance, but which are permitted explicitly for concurrent use with insured treatment under the Health Insurance Act.

There are two major categories: evaluation treatment (advanced medical care, treatment related to clinical trials, and medications approved but not yet listed in the insurance reimbursement schedule) and selected treatment (such as private room fees when hospitalized).

When utilizing the Medical Expenses Combined with Treatment Outside Insurance Coverage system, patients are responsible for the full cost of non-insured medical care. However, for medical care covered by insurance, patients only bear the portion of the price that is not covered by insurance. On the other hand, if medical care is not recognized as a medical expense combined with Treatment Outside Insurance Coverage, the insured medical care also becomes uninsured.

7.4 Expenses for Benefits for Medical Treatment (Medical Fees)

Insured medical care facilities, when providing medical care to insured patients, collect the patient's copayment at the time of service and submit a medical fee statement (referred to as a "receipt") summarizing the medical care provided over a one-month period to claim medical fees. The statement must include the frequency and points (including medication costs) for each medical procedure. An example of a medical fee statement is provided below (Fig. 7.2).

The point table used to calculate the claim amount assigns points to each medical procedure, and adjustments to medical fees (increases or decreases) are made through revisions to the point table. Revisions to the point table are proposed by the Minister of Health, Labour and Welfare after consulting the Central Social Insurance Medical Council. They are announced by ministerial ordinance based on the council's recommendations. Medical fees are calculated by adding the points assigned to each medical procedure performed, with each point valued at 10 yen. The method of paying the calculated remuneration to insured medical care facilities is called the fee-for-service payment system.

In the field of acute care hospitalization, a payment system based on a daily comprehensive evaluation using Diagnosis-Procedure Combination (DPC) has been introduced in fiscal year 2003. The diagnosis group classification, which serves as

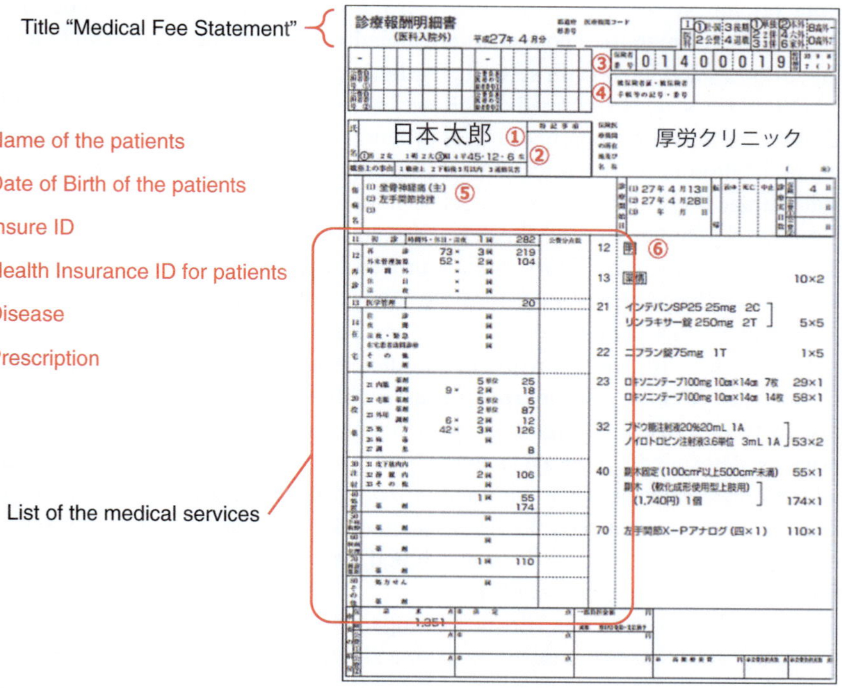

Fig. 7.2 Medical fee statement (with sample entries)

the basic unit for comprehensive payment, is composed of "disease names" based on the International Classification of Diseases and "medical procedures" such as surgeries or treatments, with over 3000 diagnosis group classifications defined. Note that medical fees have been revised approximately every 2 years since 1986.

7.5 The National Health Insurance Drug Price

Pharmaceuticals and other medical products must obtain manufacturing and sales approval under the Pharmaceuticals and Medical Devices Act before being included in the National Health Insurance (NHI) Drug Price list [5]. Pharmaceuticals must obtain manufacturing and marketing approval under the Pharmaceuticals and Medical Devices Act and be listed in the "Pharmaceutical Price Standards" to be distributed in the market [5].

Without these listings, they are not covered by insurance and are not eligible for insurance benefits. Nearly all hospitals and clinics in Japan are designated as insured medical care facilities, and pharmaceuticals prescribed in these facilities should be listed in the drug price list. Therefore, pharmaceutical companies obtain insurance coverage and a drug price listing for their products before they are brought to market [5].

The procedures for insurance coverage vary significantly depending on the type of product. For pharmaceuticals, drug prices are calculated using an independent drug pricing system which are listed in the Drug Price Standards. For medical devices, there are those whose costs are included in the medical fees paid to medical practitioners and those with separate medical fees for their use [5].

Drug prices are determined as follows: After obtaining manufacturing and marketing approval for a pharmaceutical product, the approved manufacturer submits an application to a drug price listing. The Drug Price Calculation Organization, a subordinate body of the Central Social Insurance Medical Council (CSIMC), one of the advisory bodies to the Minister of Health, Labour and Welfare, calculates the drug price, which the CSIMC then approves. After this process, the drug price is listed in the Drug Price Standards by the Minister of Health, Labour and Welfare [5].

For new drugs, drug price calculation is categorized into cases with and without similar drugs. In cases involving similar drugs, the "Similar Drug Efficacy Comparison Method" is employed; in cases without similar drugs, the "cost calculation method" is used. Foreign average price adjustments are then made, and the drug price is determined. Foreign average price adjustment is a correction made to address discrepancies when the price of a drug already exists in foreign countries, and the difference between the price calculated using a similar drug efficacy comparison method or cost calculation method and the average price of the drug in foreign countries is significant [5].

Under the Similar Drug Efficacy Comparison Method, drugs are classified into two categories: those recognized as having significant novelty and those recognized as having limited novelty. For the former, the most similar drug is selected as the comparator, while for the latter, the drug with the lowest price among similar ones over the past few years is selected as the comparator. Additionally, for the former category, adjustments may be made based on factors such as innovation, efficacy, and market potential [5].

The cost calculation method is used when there are no similar drugs available on the market, and the price is calculated by referring to manufacturing (import) costs, selling and general administrative expenses, operating profits, distribution expenses, and other relevant factors. Note that, recently, high-cost drugs and regenerative medical products have entered the market, and many of these were initially priced using the cost calculation method [5].

References

1. Kousei-roudou hakusho (in Japanese) (Ministry of Health, Labour and Welfare's White Paper); 2007. p. 4–27, ISBN-13. 978-4324083031. Data available from https://www.mhlw.go.jp/wp/hakusyo/kousei/07/dl/0101.pdf.
2. National Health Insurance Act, Act No. 192 of December 27, 1958, Japanese Law Translation. https://www.japaneselawtranslation.go.jp/ja/laws/view/2453.
3. Health Insurance Act, Act No. 70 of April 22, 1922, Japanese Law Translation. https://www.japaneselawtranslation.go.jp/ja/laws/view/3266.
4. New Essential Forensic Medicine (6th ed, in Japanese), 2019, Ishiyaku Shuppan, ISBN978-4-263-73189-5.
5. Yakka no Santei kijyun ni tsuite (Regarding the calculation standards for drug prices), Feb 14, 2024, Ministry of Health and Labour Welfare (In Japanese). https://www.mhlw.go.jp/content/12404000/001218705.pdf.

8 Long-Term Care Insurance

Jiro Ezaki, Takashi Yamashita, Shoichi Maeda, and Taisuke Izumi

In the 1960s, when Japan's Social Welfare for the Elderly was said to have begun, Japan's population aging rate, the share of the population aged 65 years and over, was 5.7%. In the 1970s, free medical care for senior citizens was implemented. However, by the 1980s, the population aging rate had reached 9.1%, and the number of long-term hospitalizations for supporting ADL or IADL that did not require medical treatment increased. By the 1990s, the aging rate had reached 12%. In 2000, the Japanese government introduced the national Long-Term Care Insurance System to shift care from institutional settings to home-based care, thereby alleviating the excessive burden of care on families, particularly those with parents and children (nuclear families). In other words, Long-Term Care Insurance in Japan can be described as a system that supports the care of senior citizens through societal efforts. In 2024, the total budget for long-term care insurance is approximately 14.2 trillion JPY, with 35,880,000 insured people aged 65 and older and 7,067,000 people (20%) requiring assistance or long-term care [1].

J. Ezaki
Ministry of Health, Labour and Welfare, The Government of Japan, Chiyoda-ku, Tokyo, Japan

Department of Biology, College of Arts & Sciences, American University, Washington, DC, USA

T. Yamashita
Department of Sociology, Anthropology, and Public Health, College of Arts, Humanities, and Social Sciences, University of Maryland, Baltimore County, Baltimore, MD, USA

S. Maeda
Graduate School of Health Management, Division of Health Care Management and Public Health, Keio University, Shinjuku-ku, Tokyo, Japan

T. Izumi (✉)
Department of Biology, College of Arts & Sciences, American University, Washington, DC, USA
e-mail: izumi@american.edu

© The Author(s), under exclusive license to Springer Nature Singapore Pte Ltd. 2026

J. Ezaki et al. (eds.), *Health Policy and Law in Japan*,
https://doi.org/10.1007/978-981-95-9241-8_8

The purposes of this Act are to improve health and medical care and to enhance the welfare of citizens. With regard to people who are under condition of need for long-term care due to disease, etc., as a result of physical or emotional changes caused by aging, and who require care such as for bathing, bodily waste elimination, and meals, and require the functional training, nursing, management of medical treatment, and other medical care, these purposes are to be accomplished by establishing a long-term care insurance system based on the principle of the cooperation of citizens, solidarity, and determining necessary matters concerning related insurance benefits, etc., in order to provide benefits pertaining to necessary health and medical services and public aid services so that these people are able to maintain dignity and an independent daily life routine according to each person's own level of abilities (Article 1 of the Long-Term Care Insurance Act; unless otherwise specified in this chapter, article numbers refer to the Long-Term Care Insurance Act) [2].

8.1 The Structure of the Long-Term Care Insurance System

Long-Term Care Insurance is a public insurance program established under the Long-Term Care Insurance Act (1997) that provides financial support to cover the costs of care for eligible adults.

The term "insurance" here means that it is a universal insurance system where all people aged 40 years and older contribute premiums, and benefits are provided to those who need them. The operational entity (insurer) is the municipality (cities, towns, and villages), which is funded by premiums and taxes. Upon enrollment in the Long-Term Care Insurance program, premiums for insured individuals aged 40 to 64 are collected alongside their universal health insurance premiums (Fig. 8.1) [2].

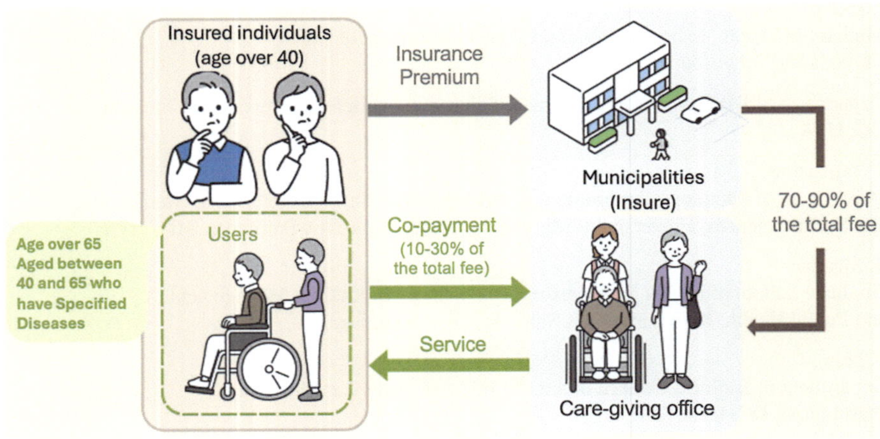

Fig. 8.1 Concept of the long-term care insurance in Japan

The methods for calculating long-term care insurance premium vary. For those enrolled in health insurance through an association or workplace health insurance, the premium is calculated by multiplying the wage by the long-term care insurance rate, with the employer bearing half of the cost. On the other hand, for insured individuals aged 65 and older, long-term care insurance premiums are generally deducted from pensions and collected by municipalities. Long-term care insurance premiums vary by municipality, depending on factors such as the availability of long-term care facilities and the number of people requiring care. The average monthly insurance premium for insured persons aged 65 and over from 2021 to 2023 was approximately 6014 JPY (average annual income for people aged 65 and over was 3,049,000 JPY in 2023).

8.2 Insured Persons

Insured persons under the long-term care insurance are divided into two categories: (i) aged 65 and older (Category 1 insured persons) and (ii) aged 40 to 64 who are enrolled in health insurance (Category 2 insured persons) [2].

For Category 1 insured individuals, Long-Term Care Insurance provides necessary insurance benefits for "a Condition of Need for Long-Term Care" or "a Needed Support Condition of the insured person," regardless of the cause. For Category 2, Long-Term Care Insurance provides necessary insurance benefits for "a Condition of Need for Long-Term Care" or "a Needed Support Condition of the insured person," and the physical or mental problems that are the causes of the Condition of Need for Long-Term Care are a result of diseases that are caused by the physical or mental changes due to aging, specified by a Cabinet Order (Specified Disease (Fig. 8.2)) [2].

When people turn 65 years old, they will receive the Certificate of Insured Person by postal mail from their municipality. For individuals aged 40 to 64, the Certificate of Insured Person is not issued. However, if an individual has a specific disease, they will receive the Certificate of Insured Person after being certified as requiring Certification of Needed Long-Term Care. Every municipal government has a long-term care insurance division or department; however, the name of the department may vary depending on the municipality (Long-Term Care Insurance Management Division, Elderly and Disabled Support Division, etc.).

1. Cancer (Terminal Stage) (Limited to cases where a physician determines, based on generally accepted medical knowledge, that the patient has reached a state with no prospect of recovery.)
2. Rheumatoid Arthritis
3. Amyotrophic Lateral Sclerosis (ALS)
4. Ossification of the Posterior Longitudinal Ligament (OPLL)
5. Osteoporosis with Fractures
6. Dementia in Early Old Age (Alzheimer's Disease, Vascular Dementia, etc.)
7. Progressive supranuclear palsy, corticobasal degeneration, and Parkinson's disease
8. Spinocerebellar degeneration
9. Spinal canal stenosis
10. Progeria (Werner syndrome)
11. Multiple system atrophy
12. Diabetic Neuropathy, Diabetic Nephropathy, and Diabetic Retinopathy
13. Cerebrovascular Disease (Cerebral Infarction, Cerebral Hemorrhage, etc.)
14. Atherosclerotic Peripheral Arterial Disease
15. Chronic Obstructive Pulmonary Disease (Emphysema, Chronic Bronchitis, etc.)
16. Osteoarthritis with Significant Deformity in Both Knee or Hip Joints

Fig. 8.2 The list of Specified Diseases for Category 2 Insured Individuals

8.3 Certification of Needed Long-Term Care

To use long-term care insurance services, individuals must first obtaina Certification of Needed Long-Term Care. This certification process is one of the significant differences from health insurance (see also Chap. 5). Once certified as "an Insured Person Requiring Long-Term Care," an individual can receive the services under the long-term care insurance (e.g., home care, home bathing care, home rehabilitation, outpatient rehabilitation, and day service) [2].

To use long-term care insurance services, eligible individuals must first apply at the long-term care insurance division in their local municipality. After that, a "Long-Term Care Support Specialist (Care Manager)" appointed by the municipality will visit the applicants' home to assess their daily living conditions (Article 27-2). Specifically, a Care Manager will assess an individual's physical functions, cognitive functions, activities of daily living, social functioning, mental and behavioral disorders, and adaptation to social life. If the individual is hospitalized, the assessment is conducted at the hospital [2].

Based on the interview/assessment results, the nationwide standardized care needs assessment software is used to determine the amount of time the applicant requires for care (initial assessment). Furthermore, based on the results of the initial assessment and the attending physician's opinion, the "Certification Committee of Needed Long-Term Care," established by the municipality, makes the final

Needed Support

Category 1: Can perform basic activities of daily living independently, but requires supervision or assistance for some tasks.
Category 2: Muscle strength has declined, making walking and standing unstable—there is a high likelihood of requiring nursing care.

Needed Long-Term Care

Category 1: Requires partial assistance with daily living, standing up, and walking. Shows a slight cognitive decline.
Category 2: Requires more care for daily living activities than Level 1, with noticeable cognitive decline.
Category 3: Requires comprehensive assistance with daily living activities. Uses a cane, walker, or wheelchair for standing and walking. Cognitive function declines, and supervision is also necessary.
Category 4: Requires assistance in all aspects of daily life, exceeding the requirements of Care Level 3. A significant decline in thinking and comprehension abilities is observed.
Category 5: Requires assistance with all daily living activities and has difficulty communicating.

Fig. 8.3 Categories of the Certification of Needed Long-Term Care and Certification of Needed Support

determination of care needs (secondary assessment). The Certification Committee of Needed Long-Term Care consists of approximately five experts in health, medical care, and welfare. Long-term care certification is generally issued within 30 days (Article 27-11) [2].

The certification results are categorized as "Certification of Needed Long-Term Care (Category 1 to 5)," "Certification of Needed Support (Category 1 or 2)," or "Not Eligible (Independent)." The municipality sends a notification letter specifying the category of Certification of Needed Long-Term Care or Certification of Needed Support (Fig. 8.3).

8.4 In-Home Service Plan and Facility Service Plan

To begin using services, a Long-Term Care Support or Care Manager must create an "In-Home Service Plan" or a "Facility Service Plan," commonly referred to as a "care plan," for a Person Requiring Long-term Care (Categories 1 to 5), which outlines when, where, and what services will be used, and for what purpose. It is crucial to finding a Care Manager who is compatible with the insured person and their family. Once the care plan is finalized, the individual and the care service provider enter a contract based on the care plan, and services are provided accordingly [2, 3].

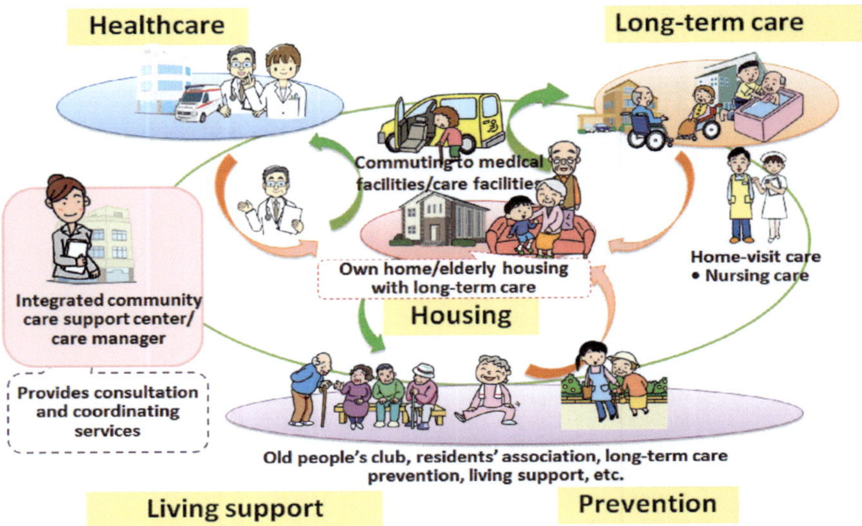

Fig. 8.4 The role of the Community General Support Center

8.5 Preventive Long-Term Care Service Plan

For a Person Requiring Support (Category 1 or 2), staff at a community general support center create a "Preventive Long-Term Care Service Plan." A Person Requiring Support can live independently with some assistance. To prevent physical decline and avoid dependence on care, they are eligible for Preventive Long-Term Care Services. These benefits include home care services, day care services, rental of specific assistive devices (e.g., walking frames and walking sticks), and home modifications (e.g., handrails and ramps installed at home). Figure 8.4 shows the idea of the Community General Support System and a community general support center [2, 3].

8.6 Long-Term Care Benefit Expense

The compensation received by care service providers is referred to as "Long-Term Care Benefit Expense." The expenses, which are set for each specific service, include the basic cost of the service plus additional or reduced amounts based on factors such as the service provider's operational capacity and the Person Requiring Long-term Care's category (i.e., the recipient's care needs). The fee schedule is determined by the Minister of Health, Labour and Welfare, after consultation with the Social Security Council (Care Allowance Subcommittee), and is reviewed every 3 years. Note that long-term care fees are based on a unit system, with factors such as the cost of living and labor costs [2, 3].

8.7 Maximum Amount of Payment

Unlike health insurance, long-term care insurance has a Maximum Amount of Payment Pertaining to an Allowance based on the level of care required (Articles 43, 55, etc.). Care managers create care plans within the payment limit. The higher the level of care needed, the higher the payment limit. If an individual or family wishes to exceed the maximum amount, they are fully responsible for the additional fees (50,320 JPY to 362,170 JPY) [2, 3].

8.8 Co-Payment

Premiums and taxes fund long-term care insurance to ensure accessibility for those who need it. The co-payment ratio ranges from 10% to 30%, depending on household income. However, as care needs increase with higher care levels, costs also rise significantly. Some households may struggle to cover these expenses solely from their pensions. The high-cost long-term care service fee system is a rule under which amounts exceeding the income-based co-payment upper limit are reimbursed. For general income levels, the upper limit is 44,000 yen per month (Article 51) [2, 3].

References

1. Kousei-roudou hakusho (in Japanese) (Ministry of Health, Labour and Welfare's White Paper); 2023. p. 41–5, ISBN-13. 978-4865793840. Data available from https://www.mhlw.go.jp/wp/hakusyo/kousei/22/dl/1-02.pdf.
2. Long-Term Care Insurance Act, Act No. 123 of December 17, 1997, Japanese Law Translation. https://www.japaneselawtranslation.go.jp/ja/laws/view/3807.
3. Kaigohoken seido no gaiyou (The overview of the Long-term case insurance) (in Japanese), Ministry of Health, Labour Welfare's webpage. Available from https://www.mhlw.go.jp/stf/seisakunitsuite/bunya/hukushi_kaigo/kaigo_koureisha/gaiyo/index.html.

Nutrition and Health

9

Jiro Ezaki, Coco Izumi, Shoichi Maeda, and Taisuke Izumi

In 2013, "Washoku" (traditional Japanese cuisine) was inscribed on the UNESCO Intangible Cultural Heritage List. In Japan, rice has had a profound influence on various aspects of the economy, society, and culture [1]. During the Edo period (1603–1868), rice served as a form of tax (yearly tribute). Rice is not only the staple food of the Japanese people but also the raw material for various seasonings such as mochi (rice cakes), sweets, miso, and vinegar, which are indispensable to Japanese cuisine. In autumn, festivals celebrating the rice harvest are held throughout the country. Rice-related foods and beverages are always served during annual events. For example, during the New Year, families eat Ozoni (a soup containing mochi, Fig. 9.1) and Osechi (New Year's dishes) while drinking sake. This close connection between year-round events where people celebrate together and Japanese cuisine is one of the reasons for the registration of rice as a UNESCO Intangible Cultural Heritage.

While agricultural mechanization has led to labor-saving measures in modern times, in the past, rice farming in rural areas was carried out with the cooperation of many villagers. This fostered a sense of community based on mutual assistance from ancient times. The recommendation based on Article 25 of the Constitution (Chap. 2) states, "The people shall also, in the spirit of social solidarity, fulfill their

J. Ezaki
Ministry of Health, Labour and Welfare, The Government of Japan, Chiyoda-ku, Tokyo, Japan

Department of Biology, College of Arts & Sciences, American University, Washington, DC, USA

C. Izumi · T. Izumi (✉)
Department of Biology, College of Arts & Sciences, American University, Washington, DC, USA
e-mail: izumi@american.edu

S. Maeda
Graduate School of Health Management, Division of Health Care Management and Public Health, Keio University, Shinjuku-ku, Tokyo, Japan

© The Author(s), under exclusive license to Springer Nature Singapore Pte Ltd. 2026

J. Ezaki et al. (eds.), *Health Policy and Law in Japan*,
https://doi.org/10.1007/978-981-95-9241-8_9

Fig. 9.1 Ozoni (Japanese New Year's soup). (From the webpage of the Ministry of Agriculture, Forestry and Fisheries. https://www.maff.go.jp/j/pr/aff/2001/spe2_03.html)

social obligations necessary for the maintenance and operation of this system, each according to their abilities." This suggests that today's social security system is rooted in Japan's cultural landscape.

9.1 Japanese Food Culture and Health

With the registration of "Washoku" as a UNESCO Intangible Cultural Heritage, Washoku has gained international recognition. Even in small towns in the United States and Europe, restaurants serving sushi and tempura have become increasingly popular, and eating Japanese food with chopsticks has become a common practice. Remarkably, Japanese food culture is gaining recognition and popularity around the world.

Today, Japan is using both old traditions and new health ideas to help people live longer and healthier lives. The basic style of Japanese cuisine, known as "Ichi-jyuu San-sai," consists of rice, soup, and three side dishes, and is said to provide an ideal nutritional balance (Fig. 9.2). "Umami" means a delicious savory taste in Japanese, and its taste is often described as the meaty, savory deliciousness that deepens flavor. The skillful use of "Umami" has enabled a diet low in animal fats, which is believed to contribute to the longevity and prevention of obesity among the Japanese population. Currently, the percentage of people in Japan with a Body Mass Index (BMI) of 25 or higher (for individuals aged 20 and above) is approximately 35% or less for men and approximately 25% or less for women. The percentage of obese individuals (BMI of 30 or higher) in Japan is significantly lower compared to other developed countries. BMI is calculated by dividing a person's weight in kilograms by their height in meters squared (kg/m^2), which estimates the amount of body fat (Fig. 9.3) [2].

On the other hand, traditional seasonings used in Japanese cuisine, such as soy sauce and miso, often contain high levels of sodium. Excessive salt intake can lead

Fig. 9.2 Example of Ichi-jyuu San-sai. (From the webpage of the Ministry of Agriculture, Forestry and Fisheries. https://www.maff.go.jp/j/yusyutu_kokusai/washoku-world-challenge/learning_01.html)

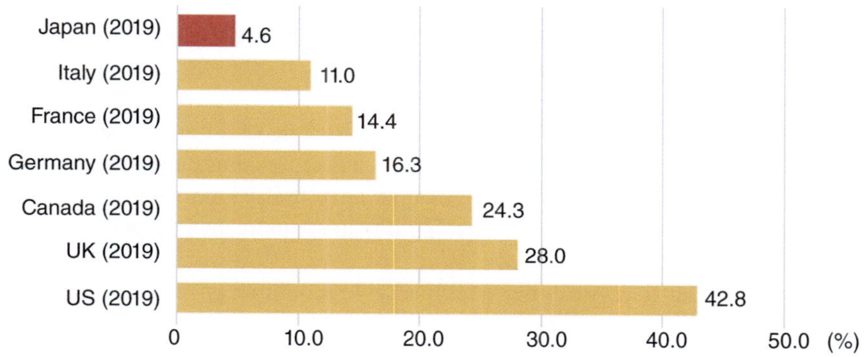

Fig. 9.3 Comparison of obesity rates (BMI 30 or higher) between Japan and developed countries. (Changes in Nutrition and Health in Japan, Ministry of Health, Labour and Welfare [2] (in Japanese). https://www.mhlw.go.jp/content/000894080.pdf)

to hypertension, which accelerates arteriosclerosis and increases the risk of life-threatening conditions such as stroke and heart disease, as well as the likelihood of requiring long-term care. For this reason, reducing salt intake is recognized as an important public health policy in Japan.

Sodium intake in Japan is relatively higher than in other countries, with consumption approximately twice the amount recommended by the World Health Organization (WHO) (less than 5 g per day). The average daily salt intake of Japanese people in 2019 was 10.1 g, showing a long-term downward trend since 1995 (13.9 g), when individual intake levels became available (Fig. 9.4) [2].

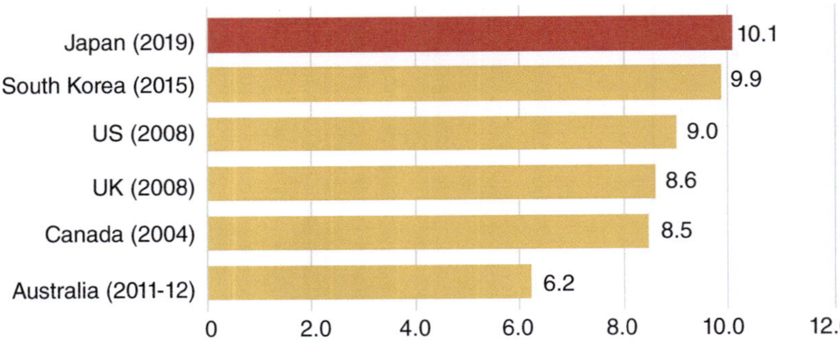

Fig. 9.4 Comparison of sodium intake in different countries. (Changes in Nutrition and Health in Japan, Ministry of Health, Labour and Welfare [2] (in Japanese). https://www.mhlw.go.jp/content/000894080.pdf)

9.2 Changes in the Nutritional and Health Status of the Japanese After World War II

In general, understanding the food culture and lifestyle of a country or region is crucial for understanding its disease structure and informing public health measures. The food culture and lifestyle of Japan have undergone significant evolution over time, with a notable impact on Japan's public health policies since World War II.

After World War II, Japan achieved remarkable economic growth, becoming the world's second-largest economy by 1968. During Japan's period of high economic growth, a change in dietary patterns known as "Westernization" of the diet occurred. As a result, the intake of grains, including rice, decreased, while the intake of meat and dairy products increased. After the war, the intake of protein and fat increased year by year. In particular, the opportunity to consume animal-based foods such as meat, dairy products, and eggs increased, leading to a rise in the proportion of animal protein in the diet (Fig. 9.5) [2].

9.3 School Lunch

One of the key policies in considering changes in the nutritional status of the Japanese population after World War II is the "school lunch program." The school lunch program originated in 1889 at the private Chua Elementary School in Tsuruoka Town (now Tsuruoka City), Yamagata Prefecture, as an initiative to encourage poor children to attend school. Furthermore, under the food shortages of the post-war period, the program was institutionalized to provide nutritional support for children.

After World War II, Japan was under the control of the Allied Forces' General Headquarters (GHQ) (also see Chap. 2 for Japan's history). All reforms in society and education were carried out under the guidance of GHQ, including the

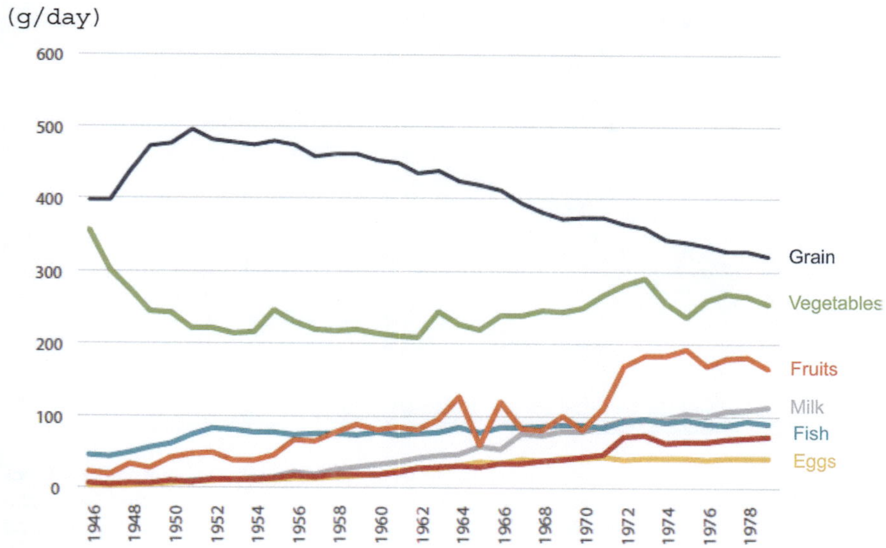

Fig. 9.5 Changes in food intake (ages 1 year and older). (Changes in Nutrition and Health in Japan, Ministry of Health, Labour and Welfare [2] (in Japanese). https://www.mhlw.go.jp/content/000894080.pdf)

introduction of school lunches. In 1946, shortly after the war, school lunches were introduced. Initially, school lunches were made using milk from UNICEF and flour from the United States. School lunches played a crucial role in helping children who had suffered from malnutrition due to food shortages during the war regain their health.

In 1954, the "School Lunch Act" was enacted, and the program expanded with government subsidies for skim milk (starting in the 1963 fiscal year) and other forms of support. By 1979, the implementation rate of school lunches in elementary schools had reached 99.4% (Fig. 9.6). The average height and weight of children during the school-age period (6–11 years old) have increased for both boys and girls (Fig. 9.7). As the implementation rate of school meals improved, the physical condition of school-age children also improved. A distinctive feature of Japanese school meals is that their purpose is not merely to enhance the nutrition of children and students but also to foster their correct understanding of food and appropriate judgment through "food education."

School Lunch Law—Article 1:

This law is established in recognition that school lunches contribute to the healthy physical and mental development of children and students and play an important role in fostering their proper understanding of food and appropriate judgment. It aims to establish necessary provisions for implementing school lunches and food-related guidance that utilizes school lunches, thereby promoting the widespread adoption and improvement of school lunches and advancing food education in schools.

Fig. 9.6 School Lunch (1947). (School lunches in Japan. In Tokyo schools, soup and vegetables are provided to Japanese children in addition to the rice they bring from home—Nagatacho National School (now Chiyoda Ward Kojimachi Elementary School). (National Archives and Records Administration, 306-NT-1152-F-16). https://search.showakan.go.jp/search/photo/detail.php?id=90016571)

9.4 Lifestyle-Related Diseases

After World War II, various measures to combat tuberculosis led to a significant decrease in the number of deaths from tuberculosis, with the mortality rate per 100,000 people falling from 216.3 in 1939 to 110.3 in 1951, approximately half the previous rate (for information on infectious disease control, see also Chap. 8). On the other hand, the mortality rates from cerebrovascular diseases, cancer (malignant neoplasms), and heart diseases have been rising annually, with cerebrovascular diseases replacing tuberculosis as the leading cause of death in 1951.

Furthermore, cancer became the second leading cause of death in 1953, and heart disease became the third leading cause in 1958, with so-called "adult diseases" dominating the top ranks of causes of death. Looking at the trend in the proportion of adult diseases among the total number of deaths in the 40s and 50s, it was 25.7% in the 40s and 37.6% in the 50s in 1947, immediately after the war, but by 1960, it had nearly doubled to 49.0% in the 40s and 62.9% in the 1950s (Fig. 9.8).

As a result, measures to combat adult diseases became a central theme of public health policy in Japan from the 1960s onwards. These adult diseases become more prevalent with age, increasing the number of people affected as the population ages. At the same time, the relationship between lifestyle habits such as smoking, diet, and exercise and these diseases has become clear, and it has been discovered that these diseases can be prevented to some extent by improving lifestyle habits. Therefore, to raise public awareness of the importance of lifestyle habits, encourage people to take the initiative in maintaining their health, and establish a system in which society supports individuals' efforts to promote health throughout their lives, the concept of "Lifestyle-related diseases" was proposed. Lifestyle-related diseases

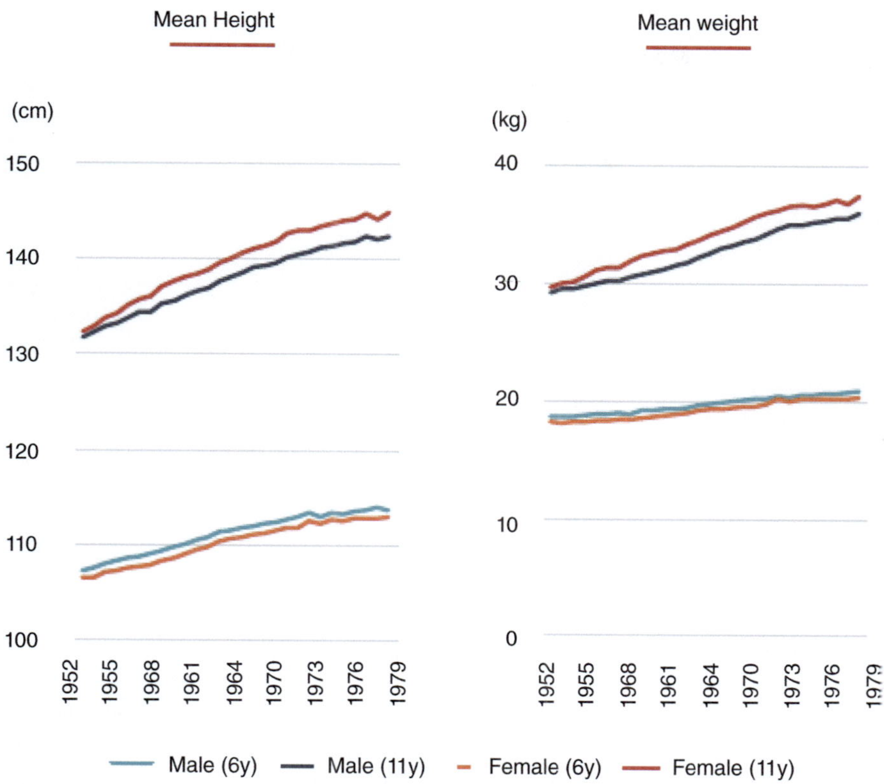

Fig. 9.7 Mean Body Height and Body Weight in 6- and 11-year-olds in Japan. (Changes in Nutrition and Health in Japan, Ministry of Health, Labour and Welfare [2] (in Japanese). https://www.mhlw.go.jp/content/000894080.pdf)

are defined as a group of diseases in which lifestyle habits, such as eating habits, exercise habits, rest, and alcohol consumption, are involved in their onset and progression.

9.5 Health Japan 21

In Japan, nationwide health promotion initiatives have been ongoing since around 1964, when the Tokyo Olympics were held. In 2000, the "Health Japan 21" initiative for national health promotion in the twenty-first century was established and is revised every 12 years. "Health Japan 21" is a national health promotion campaign promoted by the Japanese government, modeled after the US "Healthy People" initiative.

Starting in 2024, Health Japan 21 (Third Phase) will be implemented:

(i) Extending healthy life expectancy and reducing health disparities.

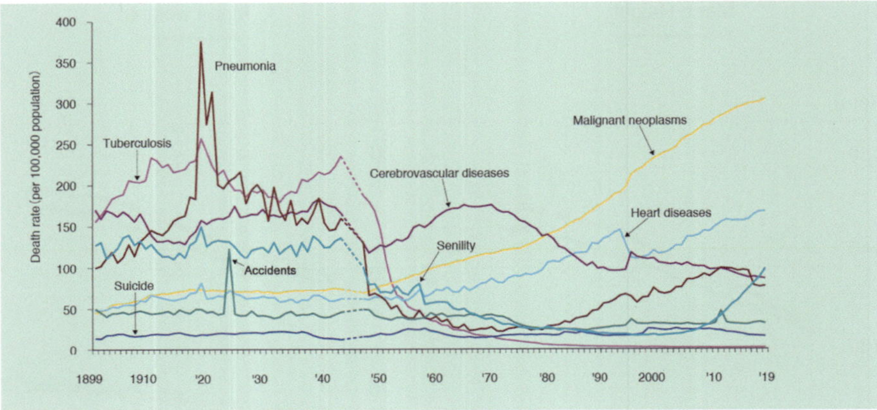

Note: For the years 1944-1946 where no graph is presented, statistics are not available because of incomplete data including loss of materials due to the damage caused by war.

Fig. 9.8 Trends in death rates from the leading cause of death (1899–2019). (https://www.mhlw.go.jp/english/database/db-hw/dl/81-1a2en.pdf)

(ii) Improving individual behavior and health status.
(iii) Enhancing the quality of the social environment.
(iv) Promoting health based on a life course approach.

These pillars will serve as the foundation for health promotion efforts. Health Japan 21 sets specific target values for items such as nutrition and diet, physical activity and exercise, rest and sleep, alcohol consumption, and smoking.

To promote Health Japan 21 and implement policies prioritizing health promotion and disease prevention, the Health Promotion Act was enacted in 2002. The Health Promotion Act stipulates national health and nutrition surveys, health guidance, nutrition management in hospitals, nursing care facilities, and school meals, as well as measures to prevent passive smoking (Fig. 9.9) [3].

9.6 Special Health Examinations and Specified Health Guidance

The percentage of Japanese men aged 20 and older with a BMI of 25 or higher had been increasing until 2007. Based on the Act on Assurance of Medical Care for Elderly People, "special health examinations" and "specified health guidance" aimed at preventing lifestyle-related diseases were mandated for insured persons and dependants aged 40–74 by health insurance providers starting in April 2008 [4]. Specific health check-ups and guidance were initiated.

Special health examinations include measurements of waist circumference, blood pressure (systolic and diastolic), lipid levels (Triglyceride, HDL (high-density lipoprotein) cholesterol), blood glucose levels, and other relevant parameters. Based on the results, individuals requiring specific lifestyle improvements are identified,

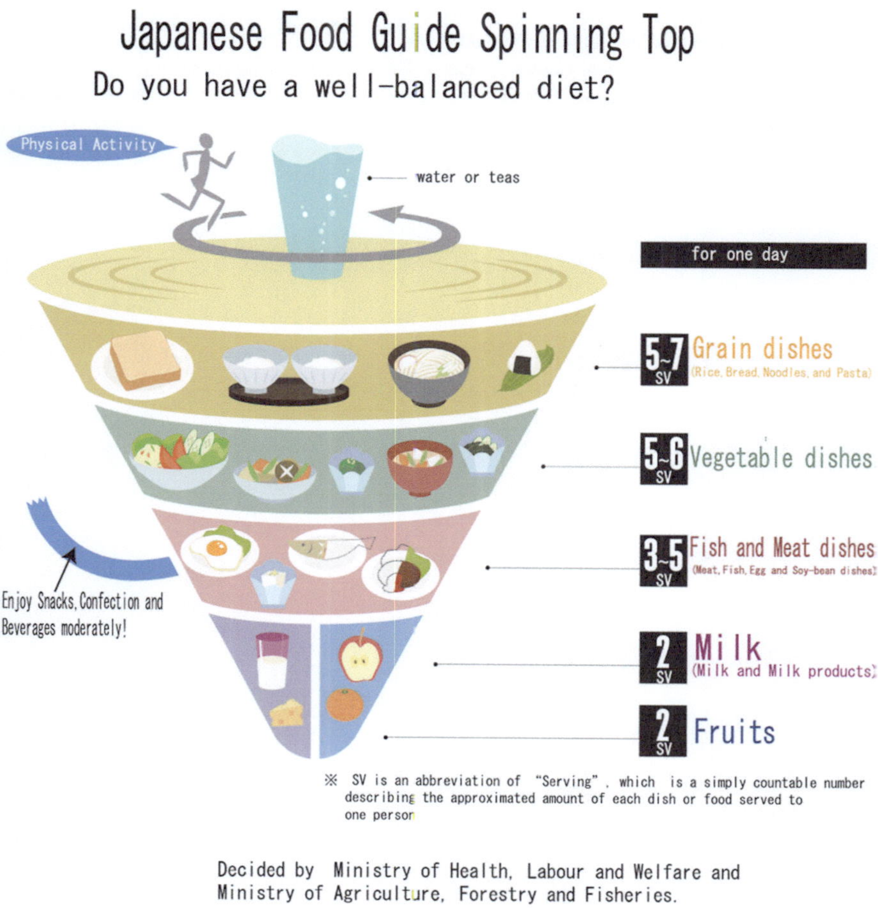

Fig. 9.9 Japanese Food Guide Spinning Top. (https://www.mhlw.go.jp/bunya/kenkou/pdf/eiyou-syokuji5.pdf)

and physicians, public health nurses, and registered dietitians provide targeted health guidance (motivational support or active support, depending on the level of risk) to help them make positive changes.

References

1. Washoku, traditional dietary cultures of the Japanese, notably for the celebration of New Year, Ministry of Agriculture, Forestry and Fisheries; 2013. https://ich.unesco.org/en/video/19697.
2. Changes in nutrition and health in Japan, Ministry of Health, Labour and Welfare; 2022 (in Japanese). https://www.mhlw.go.jp/content/000894080.pdf.

3. Health Promotion Act, Act No. 103 of August 2, 2002, Japanese law translation. https://www.japaneselawtranslation.go.jp/ja/laws/view/3727/je.
4. Act on Assurance of Medical Care for Elderly People (Tentative translation), Act No. 80 of August 17, 1982, Japanese Law Translation. https://www.japaneselawtranslation.go.jp/ja/laws/view/4903.

Public Health Organization 10

Mei Kobayashi, Hiroshi Yatsuya, Shoichi Maeda,
Taisuke Izumi, and Jiro Ezaki

The medical and nursing care systems discussed in Chaps. 3, 4, 5, 6, 7, 8, and 9, along with the public health measures introduced by the government in Chaps. 10, 11, 12, 13, 14, and 15, work together to support the health of the Japanese people. Health centers play a central role in these community health activities.

Health policies cannot succeed merely by the national government enacting laws and providing financial assistance to prefectures and municipalities; daily public health activities by public health doctors, PHNs, and other staff at prefectural and municipal governments are essential. It is fundamental and critical for PHNs at municipalities and health centers to visit households, assess the health status of each household, understand the health condition of the area, and provide guidance to prevent illness (health guidance).

Once people contract a disease, they will visit a clinic or hospital using public health insurance (Chap. 7). If they encounter problems related to social issues, they are supported by health centers and municipalities, allowing them to receive appropriate medical care, administrative assistance, and nursing care (Chap. 8).

M. Kobayashi · H. Yatsuya
Department of Public Health and Health Systems, Graduate School of Medicine, Nagoya University, Nagoya, Japan

S. Maeda
Graduate School of Health Management, Division of Health Care Management and Public Health, Keio University, Shinjuku-ku, Tokyo, Japan

T. Izumi
Department of Biology, College of Arts & Sciences, American University, Washington, DC, USA

J. Ezaki (✉)
Ministry of Health, Labour and Welfare, The Government of Japan, Chiyoda-ku, Tokyo, Japan

Department of Biology, College of Arts & Sciences, American University, Washington, DC, USA
e-mail: jezaki@american.edu

© The Author(s), under exclusive license to Springer Nature Singapore Pte Ltd. 2026

J. Ezaki et al. (eds.), *Health Policy and Law in Japan*,
https://doi.org/10.1007/978-981-95-9241-8_10

10.1 Community Health Act

The duties of health centers are stipulated in the Community Health Act as follows:

Article 6 Health centers are to plan, coordinate, and provide guidance and other necessary services for the following matters:

(i) Matters concerning the dissemination and improvement of ideas related to community health;
(ii) Matters concerning vital statistics and other statistics on community health;
(iii) Matters concerning the improvement of nutrition and food sanitation;
(iv) Matters concerning housing, water supply, sewerage, waste disposal, cleaning, and other environmental sanitation;
(v) Matters concerning medical and pharmaceutical affairs;
(vi) Matters concerning PHNs;
(vii) Matters concerning the improvement and promotion of public medical services;
(viii) Matters concerning maternal, infant, and geriatric health;
(ix) Matters concerning dental health;
(x) Matters concerning mental health;
(xi) Matters concerning the health of persons in need of long-term medical treatment due to diseases for which treatment methods have not been established or other special diseases;
(xii) Matters concerning the prevention of infectious diseases and other illnesses;
(xiii) Matters concerning sanitary tests and examinations;
(xiv) Other matters concerning the maintenance and promotion of the health of local residents.

As a result, there has been an increasing emphasis on prioritizing the perspective of individual residents who receive services. For this reason, in 1994, the Act previously known as the "Health Center Act" was revised, and its name was changed to the "Community Health Act" [1].

10.2 Activities of Health Centers

Health centers are staffed with the necessary personnel, including physicians, PHNs, dentists, pharmacists, veterinarians, radiological technologists, clinical laboratory technologists, and registered dietitians, to carry out their duties.

The head of a health center is, in principle, a physician who has at least 3 years of practical experience in public health. Health centers are established to promote and improve public health in their respective regions, and their duties are stipulated in the Community Health Act. While health centers have traditionally played a crucial role as the forefront institutions responsible for public health activities, the Ministry of Health, Labour and Welfare has been promoting the establishment of municipal health centers since 1978 to address the diversifying and sophisticated needs in the field of human health; there are approximately 2400 municipal health centers nationwide (as of 2025). Municipal health centers are facilities established to conduct essential community health activities, including health consultations, guidance, and examinations. Their establishment is stipulated in the Community Health Act. Note that health center here refers to entities originally established in 1947; however, the term public health center is also often used to distinguish these from municipal health centers, which are established by municipalities under the Community Health Act.

10.3 Public Health Nurses (PHNs)

PHNs are at the core of local government health activities. PHNs are licensed nurses who have completed an additional specialized training program in public health, usually 1 year in length, and passed a licensing examination to be qualified to work in community health settings [2]. They have clinical nursing skills, as well as the ability to promote health at the community level. PHNs work not only in government agencies but also in various other settings, including private companies and schools.

The following are examples of PHN duties at (i) prefectural health centers and (ii) municipalities.

(i) PHN activities at prefectural health centers

These include collaborating with other professionals within the organization and obtaining cooperation from municipalities and medical institutions within the jurisdiction to identify and address health issues on a wide scale. In addition, they provide regional and specialized health services for intractable diseases, tuberculosis, infectious diseases, and HIV. They establish systems for rapid and appropriate responses to health crises, including disasters, and implement pioneering health activities for new health issues, promoting their commercialization and dissemination. They also strive to actively provide technical advice and support, as well as coordination and liaison, to municipalities from a regional and specialized perspective.

(ii) Activities of municipal PHNs

As local public entities that play a fundamental role in maintaining and promoting the health of residents, municipalities are expected to address health issues that are close to residents' lives. To this end, they plan, develop, and provide health promotion and other health services in collaboration with relevant stakeholders and evaluate the effectiveness of these services. In doing so, they strive to promote a district-based system where areas within their jurisdiction are divided into several districts and PHNs are assigned responsibility for specific districts. Additionally, they participate in specific health examinations conducted by municipalities as health insurance providers. Furthermore, they formulate various health, medical, and welfare plans that reflect regional characteristics, with the participation of residents and in collaboration with relevant organizations, and implement health programs based on these plans. Moreover, they participate in the formulation of disaster prevention plans, disability plans, and urban development plans, conduct activities linked to policy implementation, and collaborate and coordinate with health, medical, welfare, and long-term care services to establish a regional care system.

10.4 Ministry of Health, Labour and Welfare, the Government of Japan

In Japan, the public health administration system is primarily structured as a unified system comprising the national government (Ministry of Health, Labour and Welfare), prefectures (with public health departments), health centers, and municipalities (with public health divisions).

Generally, health centers are established by local governments, typically at the prefectural level. However, under the Community Health Act Enforcement Ordinance, 87 health-center-designated cities (as of 2025) and the 23 special wards of Tokyo are required to establish and operate their own health centers.

The Ministry of Health, Labour and Welfare, which is responsible for general health administration and labor health administration, has 11 bureaus and two departments as internal bureaus, the Central Labor Commission as an external bureau, and various advisory councils, as well as national Hansen's disease sanatoriums, research institutes, and quarantine stations (Figs. 10.1 and 10.2). The internal bureaus of the Ministry of Health, Labour and Welfare responsible for health-related administration are as follows.

Note that the maternal and child health administration, which was previously handled by the Ministry of Health, Labour and Welfare, was transferred to the Ministry of Children and Families in 2023 to integrate it with other policies related to children. However, the two agencies continue to collaborate closely. Furthermore, the Cabinet Office for Infectious Disease Crisis Management, established in 2023, is responsible for coordinating all aspects of infectious disease crises, including preparedness, initial response, policy formulation, and interagency coordination.

Fig. 10.1 The Ministry of Health, Labour and Welfare (Tokyo, Japan) The Ministry of Health, Labour and Welfare headquarters is located in Kasumigaseki, Chiyoda Ward, Tokyo. "Kasumigaseki" is home to the headquarters of most central government ministries and agencies, including the Ministry of Health, Labour and Welfare, the Ministry of Finance, the Ministry of Foreign Affairs, the Ministry of Education, Culture, Sports, Science and Technology, and the Ministry of Land, Infrastructure, Transport and Tourism, etc

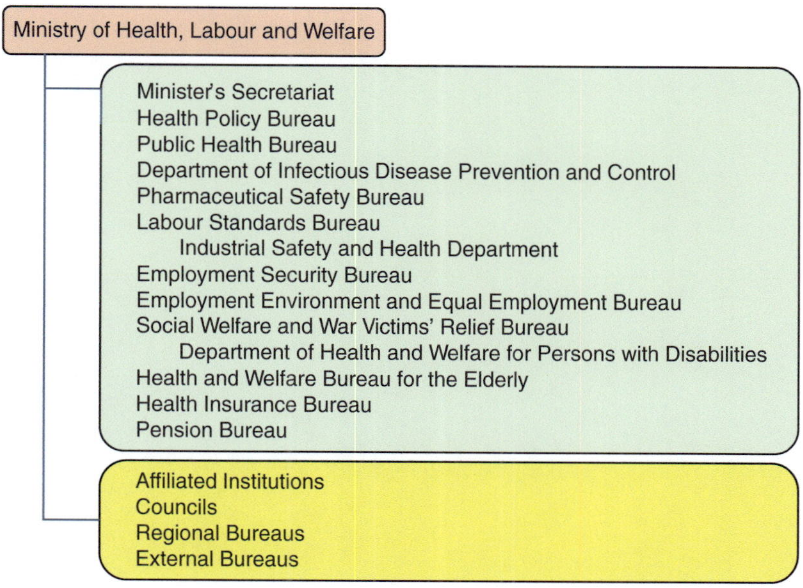

Fig. 10.2 The overview of the organization of the Ministry of Health, Labour and Welfare

References

Community Health Act, Act No. 101 of September 5, 1947, Japanese Law Translation. https://www.japaneselawtranslation.go.jp/ja/laws/view/4411.

Act on Public Health Nurses, Midwives, and Nurses, Act No. 203 of July 30, 1948, Japanese Law Translation. https://www.japaneselawtranslation.go.jp/ja/laws/view/3993.

Infectious Disease Control

11

Ayaan Zafar, Shoichi Maeda, Taisuke Izumi, and Jiro Ezaki

Approximately a century has passed since the enactment of the Infectious Diseases Prevention Act in 1897. During this period, advancements in medicine and health care, improvements in hygiene standards, and heightened public awareness of health and hygiene management have led to the elimination of infectious diseases.

However, in Japan, outbreaks of pathogenic *Escherichia coli* O-157 infections, as well as emerging infectious diseases such as Ebola hemorrhagic fever abroad, and re-emerging infectious diseases, such as malaria, are posing new threats to humanity. The Act on the Prevention of Infectious Diseases and Medical Care for Patients with Infectious Diseases, hereinafter referred to as the Infectious Diseases Control Act, was enacted in 1998. The preamble of the Act reflects the fundamental principles of public health regulations [1].

> "The Preamble of the Infectious Diseases Control Act (1998)"
>
> Mankind has over time experienced hardship due to various diseases, including infectious diseases. The spread of plagues, smallpox, cholera, and other infectious diseases has often plunged human civilization into a state of crisis. The eradication of infectious diseases has in particular been a long-sought-after goal of mankind.

A. Zafar
University of Illinois College of Medicine Chicago, Chicago, IL, USA

S. Maeda
Graduate School of Health Management, Division of Health Care Management and Public Health, Keio University, Shinjuku-ku, Tokyo, Japan

T. Izumi
Department of Biology, College of Arts & Sciences, American University, Washington, DC, USA

J. Ezaki (✉)
Ministry of Health, Labour and Welfare, The Government of Japan, Chiyoda-ku, Tokyo, Japan

Department of Biology, College of Arts & Sciences, American University, Washington, DC, USA
e-mail: jezaki@american.edu

© The Author(s), under exclusive license to Springer Nature Singapore Pte Ltd. 2026

J. Ezaki et al. (eds.), *Health Policy and Law in Japan*,
https://doi.org/10.1007/978-981-95-9241-8_11

Advances in medical science and medical care and considerable improvement of sanitation have enabled people to overcome a variety of infectious diseases. However, infectious diseases still remain a threat to humankind in new forms, through the emergence of undiscovered infectious diseases, through the resurgence of known infectious diseases, or with the development of international interpersonal exchanges.

Meanwhile, in the past in Japan there was groundless discrimination or prejudice against patients suffering from leprosy, acquired immunodeficiency syndrome (AIDS), and other infectious diseases, and those suffering from similar illnesses. The Japanese public must take these facts seriously and apply them as a moral lesson for the future.

In light of such changing circumstances surrounding infectious diseases and the situation surrounding patients of infectious diseases and other persons placed in similar situations, it is desirable to ensure high-quality and appropriate medical care for such persons and to promptly and appropriately address infectious diseases, while respecting their human rights.

From this standpoint, this Act is established to fundamentally overhaul conventional measures concerning the prevention of infectious diseases and to promote integrated measures concerning the prevention of infectious diseases and medical care for patients with infectious diseases.

This chapter introduces the Infectious Diseases Control Act. The Act places importance on respecting the human rights of patients and others, ensuring the provision of high-quality and appropriate medical care to them, and responding quickly and appropriately.

11.1 Acute Infectious Disease Control

Following the Meiji Restoration (1867), Japan actively engaged in exchanges with other countries as part of its policy of modernizing and globalizing the nation [2]. As a result, infectious diseases such as cholera were introduced from overseas. Additionally, poor sanitary conditions in cities contributed to frequent outbreaks of infectious diseases. Consequently, from the early to mid-Meiji period, the focus of public health administration was on measures against acute infectious diseases such as cholera. In the 1870s, a major cholera epidemic broke out, reaching its peak in 1879 with over 160,000 cases and more than 100,000 deaths, making it the largest cholera outbreak in Meiji history.

In response to this crisis, the government expedited the establishment of preventive measures, enacting the Temporary Regulations for the Prevention of Cholera in 1879 and the Regulations for the Prevention of Infectious Diseases in 1880. This marked the first implementation of unified and ongoing infectious disease prevention measures, helping stabilize cholera outbreaks during and after the mid-Meiji period [2].

The Infectious Disease Prevention Regulations were revised several times following its implementation. Due to significant advancements in infectious disease studies during this period, the regulations were reevaluated, leading to the enactment of the Infectious Disease Prevention Act in 1897. The Act solidified Japan's system for infectious disease prevention and, despite undergoing several amendments, remained the central law for infectious disease prevention until the enactment of the Infectious Diseases Prevention Act in 1999 [2].

11.2 Chronic Infectious Disease Control Around 1899

In the mid-Meiji period, measures against chronic infectious diseases, such as the spread of tuberculosis among urban workers, began to be addressed. In 1899, the government began compiling data on mortality statistics by cause, gradually revealing the true scope of deaths attributable to tuberculosis (Fig. 11.1) [2].

Under these circumstances, the government began to develop acts and regulations related to tuberculosis prevention, and in 1904, the Matter Concerning the Prevention of Pulmonary Tuberculosis was enacted. This was the first act in Japan concerning tuberculosis prevention [2].

However, the number of tuberculosis deaths remained above 100,000 and continued to increase until the mid-Taisho period (around 1918). In 1919, in response to growing calls for the establishment of unified tuberculosis prevention regulations, the Tuberculosis Prevention Act was enacted. The approach of the Act was similar to that of acute infectious disease prevention measures, with an emphasis on eliminating sources of infection [2].

In 1937, the Tuberculosis Prevention Act was revised, and, for the first time, a reporting system for tuberculosis patients was established. However, the reporting requirement was limited only to patients who posed a risk of transmitting tuberculosis to others in their immediate environment [2].

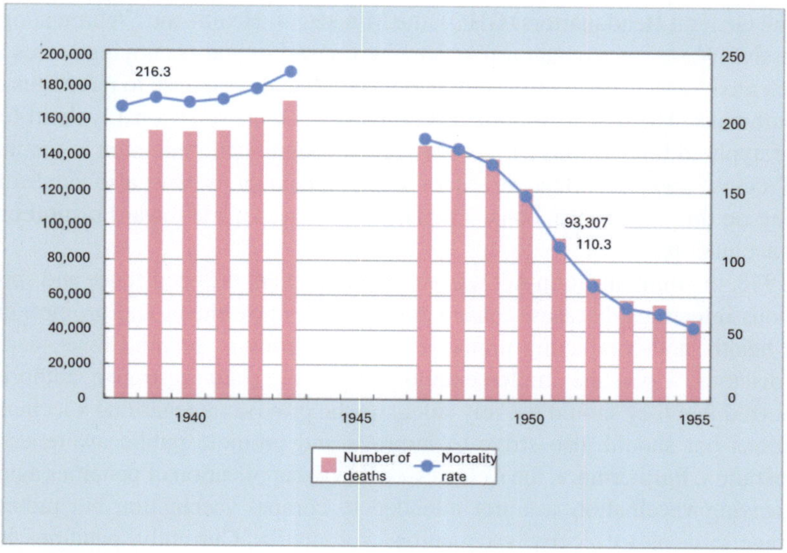

Fig. 11.1 Trends in the number of tuberculosis deaths. (Data are from Ministry of Health, Labour and Welfare's white paper (2014). https://www.mhlw.go.jp/wp/hakusyo/kousei/14/backdata/1-1-1-03.html)

11.3 The National Physical Fitness Act (1940)

In 1940, the National Physical Fitness Act was enacted to improve the physical fitness of minors and prevent tuberculosis. Physical fitness examinations were conducted annually for males aged 17–19 (and males aged 25 and under from 1942 onwards). The examination included physical measurements to assess growth and body composition, as well as disease screening with a particular focus on tuberculosis. Subsequently, the mass screening method using the tuberculin test (tuberculin reaction) and X-ray examination became the cornerstone of tuberculosis control in Japan for many years. Furthermore, these same screening methods were adopted in health examinations for factory and mine workers under labor-related laws, as well as for insured persons under the health insurance system. As a result, the number of people undergoing tuberculosis group screenings reached over 10 million people per year [2].

11.4 Infectious Disease Control After World War II (After 1945)

Immediately after the end of WW2, acute infectious diseases spread on a large scale in Japan due to the deterioration of social conditions and the successive repatriation of people from overseas. Thus, countermeasures against acute infectious diseases were the most urgent and important issues for Japan. Following strong instructions from the General Headquarters (GHQ), the Ministry of Health and Welfare adopted progressive Western vaccination practices introduced by the occupying forces and implemented widespread immunization as an emergency response to infectious disease outbreaks. The thorough implementation of vaccination against typhoid fever and paratyphoid fever from 1946 to 1947 led to a significant reduction in the number of cases, and the effectiveness of other vaccinations was also confirmed. Building on this experience, considerable progress was made toward institutionalizing vaccinations.

In 1948, the Immunization Act was enacted to prevent the occurrence and spread of various infectious diseases, contributing to the improvement and promotion of public health, and conducting immunization for pertussis, typhoid fever, and 12 other diseases. Upon the implementation of the Act, administrative authorities emphasized that they should not rely solely on the provisions regarding vaccination obligations but should also strive to improve and promote public awareness of immunization. Furthermore, the Act stressed that the application of penalties against not receiving vaccination was not intended to compel vaccination but rather to encourage citizens to receive vaccinations voluntarily. Currently, penalties have been abolished, and vaccinations are no longer mandatory, but rather an obligation of effort.

11.5 The New Tuberculosis Prevention Act (1951)

Postwar tuberculosis countermeasures underwent significant reform in response to social changes and medical advancements. In 1946, physical fitness examinations mandated by the National Physical Fitness Act were abolished, and tuberculosis screening of demobilized soldiers and repatriated individuals began in the spring of that year. Furthermore, advancements in medication for tuberculosis had a significant impact on postwar tuberculosis control measures. The most notable development was the emergence of chemotherapeutic agents such as streptomycin and calcium para-aminosalicylate (PAS). In 1950, these agents were among the first to be included in social insurance benefits and subsequently became widely used, playing a significant role in improving tuberculosis care. In 1951, a new Tuberculosis Prevention Act was enacted.

Unlike the former Tuberculosis Prevention Act enacted in 1919, which focused on preventing the spread of the disease, the new Act aimed to establish a tuberculosis prevention system focused on the progress of tuberculosis medicine during the postwar period. The new law established a series of protocols for regular health checkups, patient registration, infection prevention, patient guidance, and medical care. However, the most important aspect of the law was the incorporation of tuberculosis control measures into the social security system, which included the establishment of a system to reduce patients' medical expenses through public funds in addition to the utilization of the public medical insurance system.

11.6 Infectious Diseases Control Act (1998)

Under the previous Infectious Diseases Prevention Act, infectious diseases for which mandatory preventive measures were no longer necessary were designated as legally defined infectious diseases. However, the Act did not cover dangerous infectious diseases of global concern, such as Ebola hemorrhagic fever, and it was focused on post-outbreak responses rather than preventive measures. Additionally, there were no systematic procedural safeguards from a human rights perspective when imposing restrictions on patients' actions. These shortcomings made the Act inadequate to meet the demands of modern times. Therefore, the Infectious Diseases Control Act was enacted in 1998 [1].

Additionally, the Venereal Disease Prevention Act and the Act on the Prevention of Acquired Immunodeficiency Syndrome (AIDS) were also abolished along with the Infectious Diseases Prevention Act, and their provisions were incorporated into the Infectious Diseases Control Act. The Infectious Diseases Control Act states:

Measures for the prevention of infectious diseases shall be promoted comprehensively and systematically while taking into consideration the human rights of patients with infectious diseases, as a fundamental principle (Article 2, Infectious Diseases Control Act, the same shall apply hereinafter in this chapter). The national and local governments are obligated to take necessary measures to ensure that patients with infectious diseases receive high-quality and appropriate medical care (Article 3). Additionally, the public must ensure that the human rights of patients with infectious diseases are not infringed upon (Article 4).

In addition, the national government shall establish basic guidelines for the comprehensive promotion of infectious disease prevention and specific guidelines for infectious diseases that require particularly urgent measures. In contrast, prefectures shall establish prevention plans for the implementation of measures to prevent infectious diseases and establish provisions for the collection and disclosure of necessary information on infectious diseases [1].

Infectious diseases subject to measures under the Infectious Diseases Control Act are classified based on their infectiousness, severity when infected, and other factors (Article 6). Specifically, these include Class 1 to Class 5 infectious diseases, a novel influenza infection, designated infectious diseases, and a new infectious disease. "New infectious diseases" are classified as unknown infectious diseases deemed to pose an extremely high risk due to their infectiousness and severity upon contraction. To enable swift and appropriate responses to such diseases, measures such as hospitalization to prevent their spread are established through close collaboration between the national government and prefectures. COVID-19 has been classified as Class 5 (Fig. 11.2). Note that the classification of infectious diseases may be revised depending on the situation at the time, so readers must confirm the relevant Act and regulations each time.

Medical practitioners who diagnose patients with infectious diseases, subject to reporting (including suspected cases for all categories except Category 5 infectious diseases), must report such cases to the prefectural governor through the public health center director (Article 12) [1].

In addition, depending on the type of infectious disease, necessary measures such as disinfection of locations or objects contaminated with the infectious agent, and import quarantine for monkeys and other animals (Articles 26-3 to 36) are necessary to prevent the occurrence and spread of the disease.

Furthermore, systems and frameworks for health examinations, employment restrictions, and hospitalization must be established, and provisions are made for the burden of hospitalization costs to be shared through a combination of medical benefits under the respective medical insurance laws and public funds (Articles 37 to 41) [1].

Type of infectious disease	Key measures	Medical care system	Medical fee payment
New infectious diseases	Hospitalization, medical stay program, home convalescence	Type 1 Agreement Designated Medical Institution [The prefectural governor concludes an agreement with a designated medical institution] (enacted in April 2024)	Publicly funded in full (no insurance applied)
		Type 2 Agreement Designated Medical Institution [The prefectural governor concludes an agreement with a designated medical institution that covers outpatient care, etc.] (enacted in April 2024)	
		Designated medical institution for Specific Infections disease (several institutions nationwide designated by the government)	
Type 1 (plague, Ebola hemorrhagic fever, South American hemorrhagic fever, etc.)	Hospitalization	Designated medical institutions for Type 1 infectious disease [1 hospital in each prefecture designated by prefecture governors]	Medical insurance applied with public funds (for hospitalization)
Type 2 (avian influenza (H5N1, H7N9), tuberculosis, MERS, etc)		Designated medical institutions for Type 2 infectious disease [1 hospital in each secondary medical service area designated by prefecture governors]	
Type 3 (cholera, enterohemorrhagic Escherichia coli infection, etc.)	Work restriction in certain jobs	General medical institutions	Medical insurance applied (partial cost sharing)
Type 4 (avian influenza (excluding H5N1, H7N9), Zika virus infection, etc.)	Sterilization and other objective measures	General medical institutions	Medical insurance applied (partial cost sharing)
Type 5 (influenza (excluding avian influenza and novel influenza infection, etc.), AIDS, viral hepatitis (excluding hepatitis E and hepatitis A), new coronavirus infection (COVID-19) etc.)	Identification of the situation with infection and information (partial cost sharing) provision	General medical institutions	Medical insurance applied (partial cost sharing)
Infectious diseases such as novel influenza, etc. (novel influenza, novel coronavirus infections, etc. (excluding COVID-19))	Hospitalization, medical stay program, home convalescence	Designated medical institutions for specific/Type 1/Type 2 infectious diseases, designated medical institutions for Type 1/ Type 2 agreements	Medical insurance applied with public funds

* Infectious diseases other than Type 1, 2, or 3 infectious diseases requiring emergency measures are designated as "designated infectious diseases" in Cabinet Order and are treated the same as Type 1, 2, and 3 infectious diseases for a limited period of 1 year in principle.

Fig. 11.2 Classification of infectious diseases by the Infection Control Act

References

1. Act on the Prevention of Infectious Diseases and Medical Care for Patients with Infectious Diseases, Act No. 114 of October 2, 1998, Japanese Law Translation. https://www.japaneselawtranslation.go.jp/en/laws/view/2830.
2. Kousei-roudou hakusho (2015) (in Japanese) (Ministry of Health, Labour and Welfare's White Paper), ISBN-13: 978-4865790351. Data available from https://www.mhlw.go.jp/wp/hakusyo/kousei/15/backdata/01-00-01-001.html.

Cancer Control

12

Jiro Ezaki, Sophia Gagliardi, Shoichi Maeda, and Taisuke Izumi

Cancer has been the leading cause of death in Japan since 1981 and remains a serious issue threatening both the lives and health of the population. Thanks to progress in cancer control, mortality rates for gastric and cervical cancer have significantly decreased, and mortality and incidence rates of many other types of cancer that had previously been increasing have plateaued.

On the other hand, mortality and incidence rates for breast and prostate cancer continue to rise, and due to the aging population, the number of cancer deaths and cases are increasing for many types of cancer. Therefore, there is a strong need for the development of innovative cancer treatments and the establishment of a system that would ensure nationwide access to standard cancer care.

In 2007, the Cancer Control Act was enacted, incorporating measures to support cancer patients in the workplace and promote cancer education, thereby creating an environment conducive to further advancing cancer control efforts. The Cancer Control Act clarifies the responsibilities of relevant parties (the national government, local governments, health insurance providers, the public, and medical professionals). The Act stipulates that the national government shall establish the Cancer Control Promotion Council, an advisory body, and formulate the Basic Plan to Promote Cancer Control Programs based on the council's opinions. Based on this

J. Ezaki
Ministry of Health, Labour and Welfare, The Government of Japan, Chiyoda-ku, Tokyo, Japan

Department of Biology, College of Arts & Sciences, American University, Washington, DC, USA

S. Gagliardi · T. Izumi (✉)
Department of Biology, College of Arts & Sciences, American University, Washington, DC, USA
e-mail: izumi@american.edu

S. Maeda
Graduate School of Health Management, Division of Health Care Management and Public Health, Keio University, Shinjuku-ku, Tokyo, Japan

© The Author(s), under exclusive license to Springer Nature Singapore Pte Ltd. 2026

J. Ezaki et al. (eds.), *Health Policy and Law in Japan*,
https://doi.org/10.1007/978-981-95-9241-8_12

plan, prefectures shall develop their own prefectural cancer control promotion plans, taking into account regional characteristics. These prefectural plans are designed to comprehensively and systematically implement the fundamental policies outlined in the Act [1].

12.1 The Basic Plan for Promoting Cancer Control

The Basic Plan for Promoting Cancer Control, established in 2007, outlines the fundamental direction of cancer control efforts and has served as the basis for subsequent measures. In 2023, the Fourth Basic Plan, covering the next 6 years, was approved by the Cabinet, the chief executive body of the Japanese government, setting the overall goal of "promoting cancer control that leaves no one behind and aims to overcome cancer for all citizens" [2]. The Basic Plan is being updated every six years. It established the following three pillars: "cancer prevention," "cancer medical care," and "coexistence with cancer."

(i) Enhancing cancer prevention and screening based on scientific evidence

By increasing awareness of cancer, promoting cancer prevention, and encouraging early detection and treatment through screening, the plan aims to reduce cancer incidence and mortality rates.

(ii) Providing patient-centered and sustainable cancer care

By strengthening systems to ensure access to appropriate medical care, the plan aims to improve cancer survival rates, reduce cancer mortality rates, and enhance the quality of life for all cancer patients and their families during their treatment and care (Fig. 12.1).

(iii) Building a society where people can live with dignity and peace of mind despite cancer

By realizing a community-based society where people can live with dignity and peace of mind even after being diagnosed with cancer, we aim to improve the quality of life for all cancer patients and their families.

12.2 Promoting Cancer Prevention and Early Detection

Along with raising awareness and disseminating knowledge about cancer prevention, efforts are being made to improve the quality of cancer screening for early detection. This includes reviewing screening methods, conducting program evaluations, and ensuring training opportunities for relevant healthcare professionals. Additionally, to increase cancer screening participation rates, free tickets for cervical, breast, and colorectal cancer screenings are being implemented, along with outreach and education initiatives in collaboration with local communities and businesses.

Cancer screening in Japan is broadly categorized into "resident screening" conducted by municipalities, "occupational screening" conducted by employers or

Fig 12.1 National Cancer Center (Tokyo, Japan). (Photo by the author)

insurance providers, and “other cancer screening” (screening that individuals undergo voluntarily). While cancer screening has both benefits and risks, the types of screenings recommended by the government are those that have a proven effect in reducing cancer mortality and where the benefits outweigh the risks. The benefits of cancer screening include preventing deaths from targeted cancers.

Early detection allows for less intensive treatment, and individuals who are diagnosed as “normal” during screening can live with peace of mind knowing they do not have cancer.

On the other hand, the risks of cancer screening include false negatives, false positives, overdiagnosis, and complications. False positives can lead to unnecessary advanced testing when there is no actual cancer. In contrast, overdiagnosis, attributed to false positives, can result in unnecessary treatment for cancers that are not life-threatening. Therefore, as part of national cancer control measures, the government prioritizes screening programs with proven effectiveness in reducing cancer mortality, ensuring that individuals reach the specified age and undergo screening at appropriate intervals (Figs. 12.2 and 12.3).

	Test item	Age	Visit interval
Gastric Cancer	Clinical interview + Gastric X-ray or Gastroscope	>= 50y (Gastroscope for >= 40 y)	2 years (1 year Gastric X-ray)
Cervical Cancer	Clinical interview, inspection Cervical Cytology and internal examination	20s	2 years
		>= 30y	2 years
	Clinical interview, inspection and the primary HPV test		5 years
Lung Cancer	Clinical interview + Chest X-ray and Sputum cytology	>= 40y Sputum cytology is for >= 50 y heavy smoker	1 year
Breast Cancer	Clinical interview + Mammography	>= 40y	2 years
Colon Cancer	Clinical interview + Stool test for occult blood	>= 40y	1 years

Fig 12.2 List of cancer screenings recommended by the government

Fig 12.3 Cancer screening bus. https://www.photo-ac.com/main/detail/4462496/1?creator=aldtakasu#goog_rewarded

For screening items not recommended by the government, it is crucial to respecting employees' choices not to undergo screening from the perspective of avoiding harm. For example, CT scans, PET scans, and tumor markers are critical tests for detecting cancer recurrence or metastasis. However, their effectiveness as a cancer screening tool (i.e., a proven ability to reduce cancer mortality, with benefits outweighing risks) has not been established. Therefore, the government does not recommend these tests as part of cancer screening.

The results of cancer screenings are reported as either "no suspicion of cancer (no further testing required)" or "suspicion of cancer (further testing required)."

If "further testing required" is indicated, it means there is a suspicion of cancer, and the individual is advised to undergo further testing at a medical facility.

12.3 Promotion of Equal Access to Cancer Medical Care

The Basic Plan for Promoting Cancer Control states that the national and prefectural governments must cultivate medical professionals with specialized knowledge and skills in cancer medical care, including surgery, radiation therapy, and chemotherapy, and establish specialized cancer medical institutions. As a measure to address this, the government plans to promote the establishment of "cancer care collaboration hub hospitals" [1].

Regarding the maintenance and improvement of the quality of life for cancer patients, it is explicitly stated that palliative care will be appropriately provided according to the patient's condition to alleviate pain and suffering.

Palliative care involves multidisciplinary collaboration to provide comprehensive care, including physical symptom management and mental and psychological support, not only during the terminal stage but also from the time of diagnosis, concurrently with cancer treatment (Fig. 12.4).

Furthermore, cancer treatment in young people can affect the function of the ovaries and testicles, leading to reduced fertility, which is a significant issue for

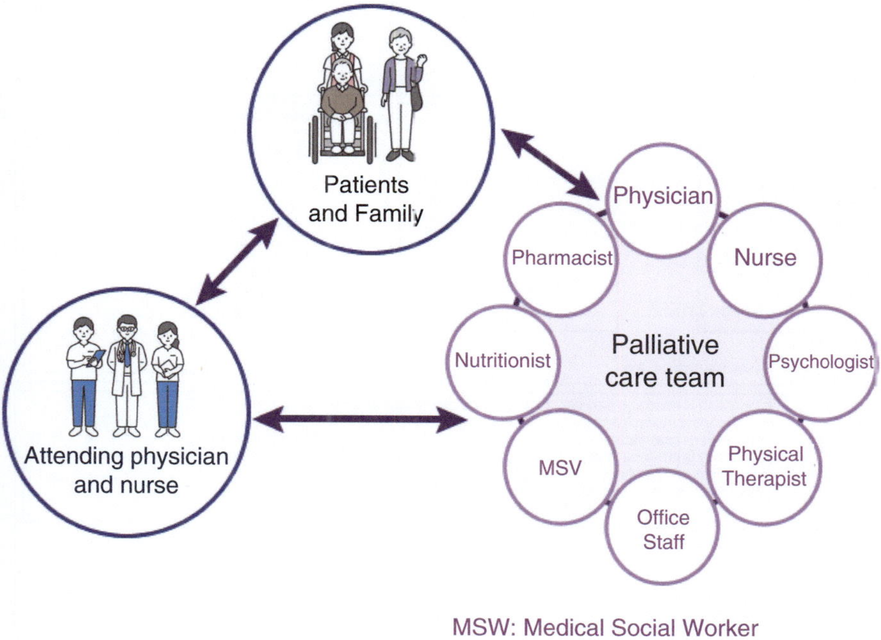

Fig 12.4 Palliative care team. MSW medical social worker

patients who wish to become pregnant. As a medical approach that prioritizes cancer treatment while supporting patients who want to have children, fertility preservation therapy is important.

On the other hand, due to the need to accumulate evidence on these treatments, a project to promote fertility preservation therapy for pediatric and Adolescent and Young Adult (AYA) cancer patients was launched in 2021 to reduce the financial burden on patients while advancing research.

12.4 Promotion of Cancer Registration

To effectively and efficiently advance cancer control measures, it is essential to grasping and analyzing the incidence and outcomes of cancer accurately. Previously, prefectures conducted regional cancer registries in accordance with the Health Promotion Act. However, with the enactment of the Act on the Promotion of Cancer Registration in 2013, the national government assumed the central role in implementing the National Cancer Registry (Fig 12.5) [1].

Concurrently, hospitals were mandated to report patient information, enabling the implementation of cancer control measures based on more accurate data. The data is also provided to researchers, the national government, and prefectural officials.

"One in two or three Japanese people will develop cancer." "The recent trend in lung cancer patients is increasing in women and non-smokers." Such information, which we often hear in our daily lives, cannot be understood without knowing the

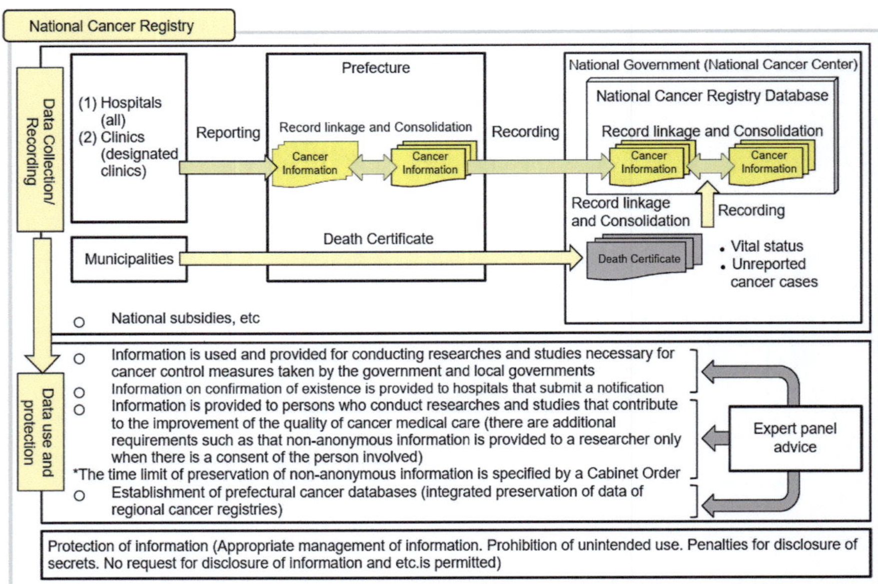

Fig 12.5 Cancer registration process

"incidence rate" of cancer, i.e., how many people are newly diagnosed with cancer each year. The only way to accurately know this number is through the cancer registration system.

The information obtained from incidence rates, such as those mentioned above, reflects the actual state of cancer and is used in national cancer control measures and prefectural regional medical plans. For example, it helps in planning and implementing measures such as determining how many cancer treatment collaboration hub hospitals should be established nationwide, how many doctors capable of treating lung cancer are needed in a particular prefecture, and what types of cancer screenings are most effective for people of specific age groups.

Cancer registration also provides various statistical information related to cancer, such as the number of cases, stage of disease, and survival rates. The stage of disease indicates how advanced the cancer is at the time of diagnosis, and by examining overall trends, it can serve as a clue to whether cancer screening is being effectively implemented at the national or prefectural level. Survival rates indicate the percentage of people diagnosed with cancer who survive afterward, and they also serve as a benchmark for how easily cancer can be cured, making them an important piece of information for doctors and patients when considering treatment options.

12.5 Supporting Cancer Research

Regarding cancer research, it is important to clarify the fundamental nature of cancer and to develop and promote innovative methods for cancer prevention, diagnosis, and treatment. Additionally, it is crucial to utilizing the results of research to reduce cancer incidence and mortality rates. From this perspective, it is also explicitly stated that the environment should be established to ensure that clinical trials for medications and medical devices with particularly high necessity are conducted swiftly and reliably, and that clinical research to develop standard treatment methods for cancer care is carried out smoothly in the Basic Plan for Promoting Cancer Control.

References

1. Cancer control Act, Act No. 98 of 2006. https://www.mhlw.go.jp/english/wp/wp-hw17/dl/02e.pdf.
2. Kousei-roudou hakusho (2024) (in Japanese) (Ministry of Health, Labour and Welfare's White Paper), ISBN 978-4-86579-438-0.

Rare Disease Measures

13

Jiro Ezaki, Dacia Marquez, Shoichi Maeda, and Taisuke Izumi

Patients with rare and intractable diseases face unique challenges compared to those with common diseases such as diabetes, hypertension, and cancer, due to the rarity of their conditions. Patients with rare and intractable diseases often undergo various tests at multiple hospitals, and it is not uncommon for them to spend many years before receiving a definitive diagnosis (diagnostic odyssey) [1].

Unfortunately, even after receiving a final definitive diagnosis, the absence of curative treatments often requires long-term care, leading to significant financial burdens. In the workplace and schools, it is difficult for patients to gain an understanding of their condition or establish patient support groups in their communities because there are typically very few others with the same disease and similar experiences. Pharmaceutical companies are also reluctant to invest in the development of new rare disease drugs, because the number of patients and the market is small. Recognizing the unique characteristics of rare and intractable diseases allows health authorities in various countries to implement different initiatives to address these conditions [1].

In Japan, a unique policy package for patients with rare and intractable diseases was established in the 1970s. The policy package comprises public funding, welfare

J. Ezaki
Ministry of Health, Labour and Welfare, The Government of Japan, Chiyoda-ku, Tokyo, Japan

Department of Biology, College of Arts & Sciences, American University, Washington, DC, USA

D. Marquez · T. Izumi (✉)
Department of Biology, College of Arts & Sciences, American University, Washington, DC, USA
e-mail: izumi@american.edu

S. Maeda
Graduate School of Health Management, Division of Health Care Management and Public Health, Keio University, Shinjuku-ku, Tokyo, Japan

© The Author(s), under exclusive license to Springer Nature Singapore Pte Ltd. 2026
J. Ezaki et al. (eds.), *Health Policy and Law in Japan*,
https://doi.org/10.1007/978-981-95-9241-8_13

services, and research support programs. In 2016, the Act on Medical Care for Patients with Intractable Diseases was enacted [1, 2].

13.1 History of the Measures for Patients with Rare and Intractable Diseases

Governmental measures involving rare and intractable diseases in Japan date back to the 1972 Outline of Measures for Intractable Diseases. Between 1950 and 1970, there was a sudden increase in the number of patients with unexplained neurological disorders (spastic paralysis of the lower limbs, deep sensory disorders, abnormal sensations, etc.) in Japan. With fatal cases being reported and the number of patients affected reaching 10,000, a widespread feeling of chaos was felt throughout Japan [1].

At first, infectious diseases were suspected to be caused by the regional clustering of these unexplained neurological disorders. However, epidemiological studies conducted nationwide revealed that the cause was an adverse drug reaction of the intestinal regulator "quinolform." The condition was later named Subacute Myelo-Optic Neuropathy (SMON).

Generally, when collecting patient data for rare diseases that occur sporadically in geographically dispersed areas, there are limitations to collecting data through individual research groups and hospitals. Additionally, when patients are transferred to other hospitals, it becomes difficult to maintain continuous data collection. One of the main reasons why the cause of SMON was elucidated in the 1970s was due to the establishment of the "SMON Research Team" by the Ministry of Health and Welfare (at the time). The SMON Research Group was a nationwide network established by neurologists, neuropathologists, and epidemiologists to identify the cause of SMON. Through the SMON Research Group, clinical data on SMON patients could be managed centrally. Additionally, patients were encouraged to participate in research by providing data. A framework was established to partially cover medical expenses through public budgets, which contributed to the continuous collection of data. Following this, numerous "rare disease research teams" modeled after the SMON Research Team were established in Japan [1].

In 1972, the Ministry of Health and Welfare (at the time) compiled the "Guidelines for Rare Disease Countermeasures," which identified three pillars: research and development, alleviating the economic burden on patients, and improving medical facilities. Over the next 40 years, nationwide efforts to address rare diseases were intensified, and the number of target diseases expanded. However, since these budgetary projects are not based on law, securing stable funding gradually became an issue [1] (Fig. 13.1).

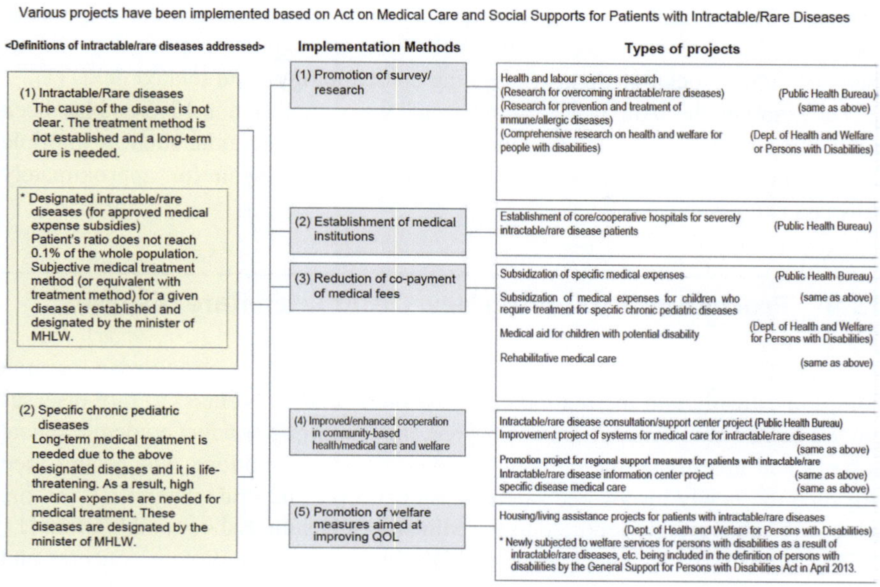

Fig. 13.1 Outline of intractable/rare disease measures

13.2 The Act on Medical Care for Patients with Intractable Diseases

In 2014, the Act on Medical Care for Patients with Intractable Diseases was enacted. The Rare Diseases Medical Care Act defines "rare diseases" as diseases that meet all four of the following criteria: (1) the mechanism of onset is unclear; (2) treatment methods have not been established; (3) the disease is rare; and (4) long-term medical care is required.

In addition to these four criteria, diseases that meet the following two criteria are designated as "designated rare diseases" by the Minister of Health, Labour and Welfare and are eligible for rare disease medical expense assistance (as described below): "5. The number of patients in Japan has not reached a certain threshold," and "6. Objective diagnostic criteria (or equivalent criteria) have been established" [1].

13.3 Rare Disease Research Groups

In Japan's efforts to address rare diseases, the "Rare Disease Research Groups" mentioned earlier play a significant role. Currently, there are more than 80 government-funded Rare Disease Research Groups, which are part of the "Ministry of Health, Labour and Welfare Scientific Research" program, a research grant program of the Ministry of Health, Labour and Welfare. Typically, a research team consists of one principal investigator and 20–30 research collaborators, operating for a period of 2–3 years. Research collaborators are experts in clinical medicine, basic medicine, social medicine (including epidemiology), and related fields [1].

For example, the "Ataxia Research Group" was established in 1975 and has been conducting long-term follow-up studies on patients with ataxia, such as multiple system atrophy and spinocerebellar degeneration, in Japan for approximately 40 years [1].

13.4 Procedures for Adding New Designated Rare Diseases

The Rare Disease Research Groups closely monitor trends in rare disease research both domestically and internationally. When new findings related to rare diseases emerge or new rare diseases are discovered, they promptly conduct epidemiological surveys and investigate the pathophysiology of these conditions. The groups then organize the newly discovered rare disease concepts and draft diagnostic criteria, which are subsequently reported to the Ministry of Health, Labour and Welfare [1].

The Ministry of Health, Labour and Welfare reviews the disease concepts and draft diagnostic criteria compiled by the Rare Disease Research Group through the Designated Rare Diseases Review Committee of the Disease Control Subcommittee of the Health Science Council, which serves as an advisory board to the Minister [1].

For example, suppose the review determines that the disease meets the six criteria for designation as a rare disease. In that case, the Ministry of Health, Labour and Welfare designates it as a new rare disease and notifies prefectures of the diagnostic criteria and diagnostic report format. As of 2025, 341 diseases have been selected as designated intractable diseases. The list of designated intractable diseases is available on the "DD rare" website [3].

13.5 Financial Assistance for Rare Disease Medical Expenses

There are 47 prefectures in Japan. All prefectures are required to designate physicians who meet specific criteria, such as having at least 5 years of experience in treating rare diseases and holding a specialty certification from a relevant medical society, as "designated rare disease physicians" [2].

Designated rare disease physicians conduct tests to determine whether patients with suspected designated rare diseases meet the diagnostic criteria. Suppose they determine that the requirements for a designated rare disease are met. In that case, they complete the diagnostic form given by the Ministry of Health, Labour and Welfare and issue it to the patient. The diagnostic form includes the patient's name, date of birth, and clinical findings (physical examination results, blood test data, imaging findings, and genetic test results if applicable) that are clinically or research-relevant for each designated rare disease. The patient then applies to the prefecture for financial assistance with medical expenses for a rare disease. At this time, the

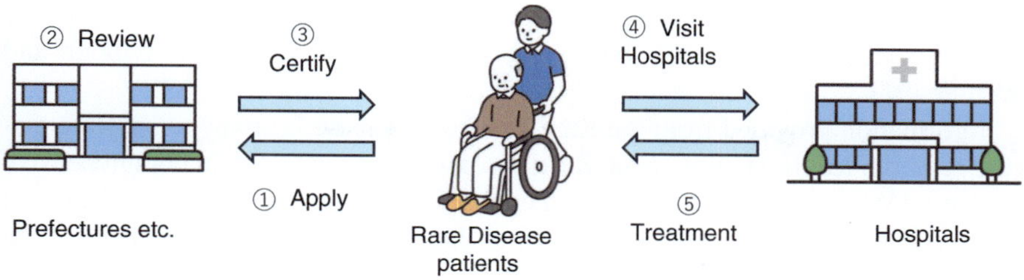

Fig. 13.2 How do the patients receive medical expense assistance for designated intractable diseases?

applicant (patient) must submit the application form along with the diagnostic report issued by the physician [2].

Each prefecture has a "Designated Rare Disease Review Committee," composed of clinical experts in rare disease medical practice in the region, typically university professors or department heads of major hospitals. They then review the documents submitted by the patients. If the criteria for a designated rare disease are met, the prefecture issues a "Rare Disease Medical Expense Recipient Certificate." The validity period of the certificate is 1 year. Therefore, if a patient with a rare disease wishes to continue receiving medical expense assistance, they must obtain a diagnostic report from a designated physician specializing in rare diseases annually and submit a renewal application (Fig. 13.2).

Japan has a universal health insurance system, so in principle, all citizens are enrolled in health insurance (see also Chap. 7). In principle, 70% of medical expenses are covered by the insurance provider, and the patient bears the remaining 30%. Patients with rare diseases can present their health insurance card along with the "Rare Disease Medical Expense Recipient Certificate" at the hospital's front desk to have their medical expenses reduced or waived based on their income. The reduced or waived portion constitutes the medical expenses for rare diseases, which are shared equally between the national government and the prefectural government.

In Japan, it is possible to receive treatment for rare diseases at local hospitals and clinics. However, prefectures designate hospitals with particularly well-developed medical care systems for rare diseases (mostly university hospitals or public hospitals) as "rare disease medical care coordination hub hospitals."

These hub hospitals conduct examinations for patients with undiagnosed or difficult-to-treat conditions, refer them to more specialized medical institutions, and consult with rare disease research teams.

13.6 Rare Disease Database

All diagnostic reports submitted by patients with rare diseases to their prefectures are sent annually to the Ministry of Health, Labour and Welfare. It is then converted into electronic data and stored in a centralized database (Rare Disease Database).

Access to the Rare Disease Database is granted to administrative agencies and rare disease research teams after review by the Ministry of Health, Labour and Welfare [2].

Information provided from the Rare Disease Database is anonymous. Rare disease research teams can utilize the Rare Disease Database to obtain comprehensive epidemiological data on patients with the rare diseases they are studying, which is a distinctive feature of rare disease research in Japan. Additionally, information from the Designated Rare Disease Patient Database may also be provided to private-sector entities conducting research and development in the medical or welfare fields related to rare diseases.

13.7 Promotion and Awareness Raising of Intractable Disease Countermeasures

The concerns of patients with intractable diseases extend beyond medical care to welfare, employment, and education. Prefectures have established "Intractable Disease Consultation Support Centers" to provide consultation on medical and social issues. Some centers are located within medical institutions and staffed by medical social workers or nurses, while others are operated primarily by patient organizations.

Additionally, the Public Interest Incorporated Foundation for Rare Disease Medical Research (Tokyo) operates the "Rare Disease Information Center" with assistance from the Ministry of Health, Labour and Welfare. These centers provide information necessary for medical treatments, daily life (disease explanations, diagnoses, treatment guidelines, etc.), national measures for rare diseases, overviews of various systems, research groups on intractable diseases, patient groups, patient families, and medical professionals via the Internet. With over 4 million visits per month, these centers have become an indispensable information tool for those involved in rare disease measures in Japan.

References

1. Ezaki J, et al. Japanese Public Health Insurance System's new genomic strategic action to shorten the "diagnostic odyssey" for patients with rare and intractable diseases. J Hum Genet. 2024;69(11):549–52.
2. Act on Medical Care for Patients with Intractable Diseases, Act No. 50 of May 30, 2014. https://www.japaneselawtranslation.go.jp/ja/laws/view/3363.
3. DDrare. Information about drug development for rare and intractable diseases. National Institutes of Biomedical Innovation, Health and Nutrition (NIBN). https://ddrare.nibn.go.jp/index_e.html.

Maternal and Child Health 14

Eri Ishikawa, Ayaan Zafar, Shoichi Maeda, Taisuke Izumi, and Jiro Ezaki

Maternal and child health measures, including health guidance for pregnant women and infants, were traditionally regarded as part of child welfare administration.

Supplementary Information The online version contains supplementary material available at https://doi.org/10.1007/978-981-95-9241-8_14.

E. Ishikawa
Graduate School of Health Management, Division of Health Care Management and Public Health, Keio University, Shinjuku-ku, Tokyo, Japan

A. Zafar
University of Illinois College of Medicine Chicago, Chicago, IL, USA

S. Maeda
Graduate School of Health Management, Division of Health Care Management and Public Health, Keio University, Shinjuku-ku, Tokyo, Japan

T. Izumi
Department of Biology, College of Arts & Sciences, American University, Washington, DC, USA

J. Ezaki (✉)
Ministry of Health, Labour and Welfare, The Government of Japan, Chiyoda-ku, Tokyo, Japan

Department of Biology, College of Arts & Sciences, American University, Washington, DC, USA
e-mail: jezaki@american.edu

© The Author(s), under exclusive license to Springer Nature Singapore Pte Ltd. 2026

J. Ezaki et al. (eds.), *Health Policy and Law in Japan*,
https://doi.org/10.1007/978-981-95-9241-8_14

However, in 1965, Japan's maternal mortality rate was higher than that of other advanced countries. Despite significant improvements in infant mortality rates, maternal mortality rates, physical condition, and nutritional status following World War II, regional disparities in these areas persisted (Fig. 14.1). Considering this situation, the Maternal and Child Health Act was enacted in 1965 [1].

The Maternal and Child Health Act establishes the principles of protecting and respecting motherhood, as well as maintaining and promoting the health of mothers and their children. It clarifies the roles of mothers, infant guardians, and administrative agencies. Specifically, mothers are obligated to deepen their understanding of pregnancy, childbirth, and childcare, and to strive to maintain and promote their health. Guardians of infants and young children are obligated to deepen their understanding of childcare and to strive to preserve and promote the health of infants and young children (Maternal and Child Health Act, Article 4) [1].

14.1 Maternal and Child Health Handbook

Pregnant women must promptly notify the mayor of their municipality of their pregnancy (Maternal and Child Health Act, Article 15). Municipalities must issue a Maternal and Child Health Handbook to those who have reported their pregnancy (Maternal and Child Health Act, Article 16). Generally, municipalities issue a

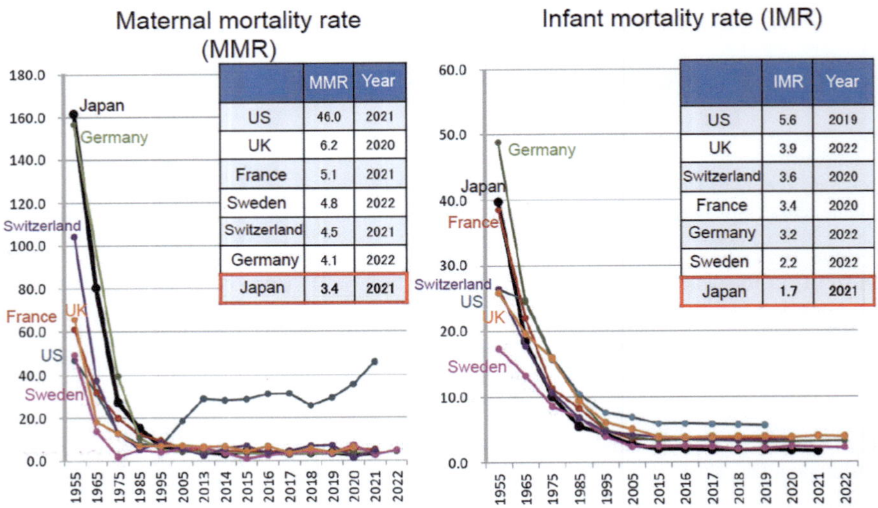

Fig. 14.1 Maternal and Infant mortality rate. The Maternal Mortality Ratio (MMR) is the number of maternal deaths (deaths related to pregnancy/childbirth) per 100,000 live births, covering deaths during pregnancy, childbirth, and up to 42 days after pregnancy ends, excluding accidents. The infant mortality rate (IMR) is the number of deaths of infants under one year old, expressed per 1000 live births in a specific area and time period. https://www.cfa.go.jp/assets/contents/node/basic_page/field_ref_resources/1ceca2fc-2bfe-4657-bf45-ac8aec94171e/d43b4a43/20250312_councils_shingikai_seiiku_iryou_1ceca2fc_07.pdf

Maternal Health Examination Voucher along with the Maternal and Child Health Handbook (Fig. 14.2) [1].

Furthermore, pregnant women and new mothers must record relevant information in the Maternal and Child Health Handbook each time they receive a health examination or health guidance from a physician, dentist, midwife, or public health nurse. The same applies to the guardians of infants or toddlers who have received health examinations or health guidance (Maternal and Child Health Act, Article 16) [1].

The Maternal and Child Health Handbook is a comprehensive health record for pregnant women, infants, and toddlers, documenting records of health examinations, health guidance, and vaccination status. Therefore, even when maternal and child health services are provided by different professionals at different locations and times, continuous and consistent care is possible. Furthermore, the Maternal and Child Health Handbook provides information on administrative services, health, and childcare from pregnancy through infancy [1].

Health guidance on pregnancy, childbirth, and childcare is primarily provided by municipalities (Maternal and Child Health Act, Article 10). For pregnant women, newborns, and premature infants, doctors, midwives, or public health nurses visit their homes as needed to provide health guidance (Maternal and Child Health Act, Articles 11, 17, and 19). To promote the health of pregnant women and their

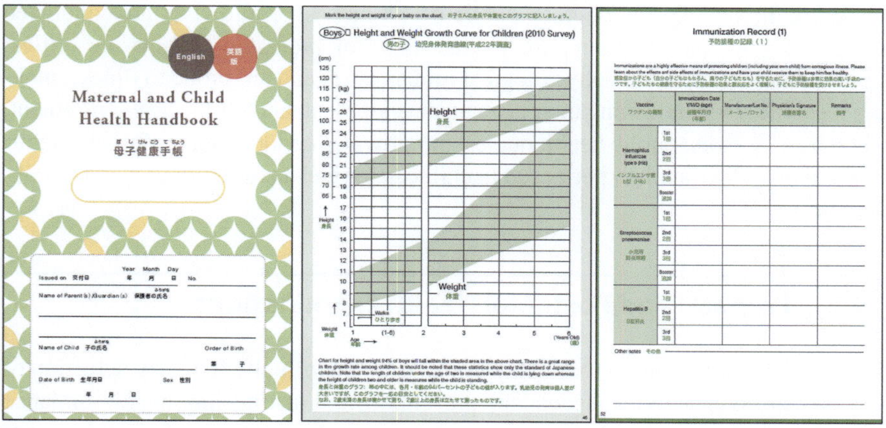

Fig. 14.2 Maternal and Child Health Handbook. The Maternal and Child Health Handbook (Boshi Kenkō Techo) is a booklet issued by municipalities to pregnant women who have registered their pregnancy, in accordance with Article 16 of the Maternal and Child Health Act. It records health management information for the mother during pregnancy and for the child after birth. It is commonly known simply as the Maternal and Child Handbook. Some local governments use the names Parent-Child Handbook or Parent-Child Health Handbook. It serves as a foundational resource for health guidance for pregnant women, new mothers, and infants, while also functioning as a parenting guide for the child's guardians. The first part of the handbook format contains sections for recording pregnancy and childbirth. For the child after birth, it includes sections for recording regular health checkups through elementary school enrollment, vaccinations (such as diphtheria, pertussis, tetanus, polio, and measles), dental examinations, and other relevant information. The latter part of the handbook is designed with content tailored to each municipality's specific characteristics. https://mchbook.cfa.go.jp/pdf/R5_EN_240301.pdf

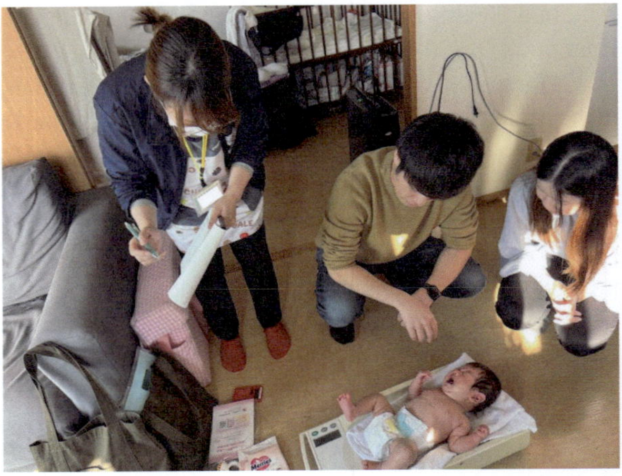

Fig. 14.3 Public health nurses, midwives, and other professionals from municipalities visit households with newborns to measure the baby's weight, assess the baby's health, discuss parenting concerns and the mother's physical condition, and introduce childcare services. (Photo from Shimada City, Shizuoka, Japan)

families, guidance and advice are provided not only on health and hygiene but also on all aspects of daily life, taking into account the family and living environment (Video 14.1, Fig. 14.3).

Newborns have weak resistance to the outside world and require special attention to nutrition, environment, and disease prevention. When deemed necessary for childcare, home visits by doctors, public health nurses, or midwives are conducted to provide guidance and support. Home visits for premature infants are critical because they have various physiological immaturities and are prone to illness, making it essential to confirming their growth, development, and nurturing environment. Therefore, doctors, public health nurses, or midwives conduct home visits for families where parents are inexperienced in childcare or feel anxious about it and require support [1, 2].

14.2 Prenatal and Postnatal Health Examinations

Health examinations are important not only as an opportunity for early detection of diseases or abnormalities (secondary prevention) but also to link health guidance aimed at preventing disease occurrence through early detection of risks (primary prevention). Pregnant women and infants can receive health checkups by methods established by municipalities, and detailed examinations are conducted as necessary. In recent years, due to factors such as the rising age of childbirth, the number of pregnant women requiring enhanced health management has been increasing. Additionally, some pregnant women do not undergo health checkups due to economic reasons. Therefore, the importance and necessity of health guidance and health checkups for pregnant women have become even more critical in maintaining and promoting the health of both mothers and fetuses. Regarding maternal and child health examinations, since 2010, HTLV-1 antibody testing has been included as a standard examination item in maternal health examinations, and since 2011, genital chlamydia testing has also been added.

Regarding postpartum health examinations, the importance of health examinations for women shortly after childbirth has been emphasized, not only for the recovery of the mother's physical condition and assessment of her mental state, but also from the perspective of preventing postpartum depression and abuse of newborns. Therefore, starting in 2017, a program was launched to subsidize the costs of two postpartum health examinations conducted by municipalities, aiming to strengthen support for mothers and infants in the early postpartum period and establish a seamless support system from pregnancy through the child-rearing period.

14.3 Infant Health Checkups

At 18 months of age, developmental milestones such as walking and language skills become easier to observe, so municipalities conduct health checkups for children aged 18 months. This checkup focuses on the early detection of physical and mental disabilities, the prevention of dental caries, and the assessment of nutritional status, while also providing guidance to parents on nutrition, psychology, and childcare. Furthermore, for those who require further examinations or follow-up guidance based on the results of the health examination, appropriate measures are taken in close collaboration with the individual's primary care physician or other relevant professionals.

Additionally, at the age of 3 years, a health examination is conducted by municipalities to detect early signs of physical growth, mental development, and visual or auditory impairments. Similar to the 18-month health examinations, further examinations are conducted as needed.

14.4 House-Call Services for All Households with Infants

Traditionally, doctors, midwives, and public health nurses have conducted home visits for newborns and premature infants as needed to provide health guidance. Since 2009, in addition to these conventional visits, a new initiative, "house-call services for all households with infants," was launched to prevent isolation in child-rearing. Under this initiative, home visits are conducted for all households with infants aged up to 4 months (Child Welfare Act, Article 6-3) [3].

This program includes the following activities:

(i) Listening to and providing consultation for any concerns related to childcare.
(ii) Providing information on child-rearing support.
(iii) Assessing the physical and mental condition of the infant and their caregivers, as well as the overall caregiving environment.
(iv) Considering and coordinating appropriate services and support for families in need, in collaboration with relevant organizations.

In addition to public health nurses, midwives, and nurses, personnel such as nursery schoolteachers, maternal and child health promoters, child welfare volunteers, and experienced parents may also be involved in conducting these visits. These individuals may be required to complete training programs, depending on necessity.

14.5 Neonatal Care for Premature Infants

Premature infants are physiologically immature compared to normal newborns, are more susceptible to illness, and may develop physical or mental disabilities. Therefore, appropriate measures must be taken promptly after birth. As a result, medical care necessary for the care of premature infants is provided, and municipal public health nurses or other professionals may conduct home visits to guide the parents of premature infants as needed. When an infant weighing less than 2500 grams is born, the parents must promptly notify the municipality where the infant is currently located (Maternal and Child Health Act, Article 18). Additionally, municipalities provide subsidies for the portion of medical insurance costs that parents must bear for premature infants requiring hospitalization at a hospital or clinic for care (Maternal and Child Health Act, Article 19).

14.6 Newborn Screening

Phenylketonuria (PKU) and other congenital metabolic disorders, as well as congenital hypothyroidism, can be prevented from causing intellectual disabilities or other physical and mental impairments through early detection and treatment. Therefore, newborn mass screening tests using blood are conducted for all newborns in prefectures and designated cities. Newborn mass screening is undertaken in all municipalities and has a high participation rate.

Additionally, hearing impairments can have minimal impact on speech and language development if detected early and appropriate support is provided. Therefore, municipalities also conduct newborn hearing tests.

14.7 Municipal Child and Family Centers

Municipalities have established Child and Family Centers to provide integrated consultation and support services to all pregnant women, families with children, and children. These centers offer support services, including pregnancy registration, assistance for pregnant women, and consultation on child-rearing and child-related issues, and develop support plans to connect families with suitable services (Maternal and Child Health Act, Article 22).

Many local governments offer services such as prenatal classes, baby food (weaning) classes, parenting consultations, and toothbrushing classes (Maternal and

Child Health Act, Articles 9 and 10). In addition, child-rearing support spaces such as "Kosodate Hiroba" (child-rearing plazas) and parenting salons are also available as places for local parents and children to interact (Child Welfare Act, Article 6-3).

References

1. Maternal and Child Health Act, Act No. 141 of August 18, 1965.
2. New Essential Forensic Medicine (6th edition, in Japanese), 2019, Ishiyaku Shuppan, ISBN 978-4-263-73189-5.
3. Child Welfare Act, Act No. 164 of December 12, 1947, Japanese Law Translation. https://www.japaneselawtranslation.go.jp/ja/laws/view/4035.

Mental Health and Welfare 15

Jiro Ezaki, Tristan Hotchkin, Shoichi Maeda, and Taisuke Izumi

In Japan, matters related to mental health and welfare are stipulated in the Act on Mental Health and Welfare for Persons with Mental Disorders or Disabilities (The Mental Health and Welfare Act) [1].

The purpose of this Act is to promote the welfare of persons with mental disorders or disabilities and improve the mental health of citizens, by providing for the medical care and protection of persons with mental disorders or disabilities; by providing necessary aid to promote the social reintegration, independence, and participation in socio-economic activities of persons with mental disorders or disabilities, in combination with the Act on Providing Comprehensive Support for the Daily Life and Life in Society of Persons with Disabilities (Act No. 123 of 2005); and by preventing mental disorders or disabilities and maintaining and improving the mental health of citizens (Mental Health and Welfare Act, Article 1) [1].

Until 1875, there were no legal regulations regarding mental health. However, in 1875, the first public hospital for mental disease was established in Japan (initially called "Tenkyouinn"). At this time, psychiatry began to be taught in medical schools, laying the groundwork for the development of modern psychiatry in Japan. In 1900, the "Mental Patients Care Act" was enacted, designating spouses, parents, and other

J. Ezaki
Ministry of Health, Labour and Welfare, The Government of Japan,
Chiyoda-ku, Tokyo, Japan

Department of Biology, College of Arts & Sciences, American University, Washington, DC, USA

T. Hotchkin · T. Izumi (✉)
Department of Biology, College of Arts & Sciences, American University,
Washington, DC, USA
e-mail: izumi@american.edu

S. Maeda
Graduate School of Health Management, Division of Health Care Management and Public Health, Keio University, Shinjuku-ku, Tokyo, Japan

© The Author(s), under exclusive license to Springer Nature Singapore Pte Ltd. 2026

J. Ezaki et al. (eds.), *Health Policy and Law in Japan*,
https://doi.org/10.1007/978-981-95-9241-8_15

relatives as guardians responsible for the care of a person with mental disorders. However, "private confinement" was widely practiced [1].

Following World War II, the establishment of the Japanese Constitution designated the improvement of public health as a national responsibility. In 1950, the Mental Health Law was enacted to provide appropriate medical care and protection for people with mental disorders.

In 1964, an incident occurred in which U.S. Ambassador to Japan Edwin O. Reischauer (1910–1990) was stabbed and injured by a teenager with schizophrenia, highlighting the inadequate medical care available to people with mental disorders and making it a major social issue. In 1965, the Mental Hygiene Act was revised to establish a system of public assistance for outpatient care and to designate public health centers as the frontline institutions for mental health administration in local communities.

Additionally, following a human rights abuse incident at a psychiatric hospital in 1983 (Utsunomiya Hospital Incident), calls grew for the protection of the human rights of mental health patients, including those hospitalized. In 1987, the Mental Hygiene Act was revised to ensure the provision of appropriate medical care and protection for individuals with mental health issues, while promoting their social reintegration. The revisions included the establishment of a voluntary hospitalization system and a mental health review board; the Act was subsequently renamed the Mental Health Act [1].

In 1993, the Basic Act for Persons with Disabilities was enacted, clearly designating people with mental disabilities as subjects of the Act. In 1995, the Mental Health Act was renamed the Mental Health and Welfare Act. The Act was revised to incorporate welfare elements aimed at promoting self-reliance and social participation, while also strengthening welfare measures for persons with mental disabilities, in addition to existing health and medical policies.

In the 2022 revision of the Mental Health and Welfare Act, systems were established for reporting abuse of patients by staff at psychiatric hospitals, including reports by those who discover abuse and notifications from patients (Mental Health and Welfare Act, Article 40-3). Persons with mental disorders covered by the Act include those with schizophrenia, acute intoxication or dependence on psychoactive substances, intellectual disabilities, psychopathic disorders, and other mental disorders (Mental Health and Welfare Act, Article 5) [1, 2].

15.1 Designated Mental Health Physicians

The Minister of Health, Labour and Welfare designates individuals with the necessary knowledge and skills to determine the necessity of compulsory hospitalization or medical protection hospitalization, restrictions on behavior, etc., based on applications, as "Designated Mental Health Physicians" (Mental Health and Welfare Act, Article 18). Designated mental health physicians are physicians who have at least 5 years of clinical experience, at least 3 years of experience in psychiatric treatment, have completed the required training, and have had their case reports approved by the Minister of Health, Labour and Welfare. They are responsible for determining the necessity of involuntary hospitalization and restrictions on the behavior of hospitalized patients, among other important duties related to the protection of the human rights of hospitalized patients [1, 2].

15.2 Medical Care and Protection, and Types of Hospitalization

The following forms of hospitalization are specified for individuals with mental disorders admitted to psychiatric hospitals:

(i) Voluntary hospitalization (Mental Health and Welfare Act, Article 20):
Hospitalization based on the individual's consent.

(ii) Involuntary Hospitalization by a Prefectural Governor (Mental Health and Welfare Act, Article 29):
Hospitalization is carried out when, based on a report or notification from a police officer or other authorized person, two or more designated mental health physicians designated by the governor of the prefecture or other relevant authority examine the individual and determine that the individual is a person with a mental disorder and that hospitalization is necessary for medical treatment and protection, as well as a risk that the individual may harm themselves or others due to their mental disorder, and the results of the examinations by the designated physicians are consistent.

(iii) Emergency Involuntary Hospitalization by a Prefectural Governor (Mental Health and Welfare Act, Article 29-2):
Admission may be made for up to 72 hours in cases where urgent action is required. The procedures for admission by order cannot be followed, based on the results of an examination conducted by a designated mental health physician designated by the prefectural governor or equivalent authority, who determines that the individual is a person with mental illness and that there is a significant risk of self-harm or harm to others if admission is not immediately granted.

(iv) Involuntary Hospitalization for Medical Care and Protection (Mental Health and Welfare Act, Article 33):
When, as a result of an examination by a designated mental health physician, a person is determined to be a person with a mental disorder and requires hospitalization for medical treatment and protection, and it is determined that the person is not in a state where hospitalization under the provisions of Article 20 (voluntary hospitalization) can be carried out, hospitalization may be carried out with the consent of the person's family members (spouse, legal guardian, or guardian) or, if there are no such family members or if all family members are unable to express their consent, with the consent of the mayor of the municipality where the person resides. Or, if the individual has no family members or if all family members are unable to

express their consent, and with the consent of the mayor of the municipality where the individual resides, hospitalization may be conducted without the individual's consent (provisions regarding measures to promote transition to community living after discharge from a psychiatric hospital are stipulated in Articles 33-4 to 33-6 of the Act).

(v) Emergency hospitalization (Mental Health and Welfare Act, Article 33-7):

In cases where immediate action is required and consent from family members or others cannot be obtained, if, based on the examination by a designated mental health physician, the individual is determined to be a person with a mental disorder who, if not hospitalized immediately, would face significant obstacles in receiving medical care and protection, and who is not in a state where hospitalization under the provisions of Article 20 (voluntary hospitalization) can be carried out, hospitalization may be conducted without the individual's consent for a maximum of 72 hours.

The administrator of a psychiatric hospital may impose necessary restrictions on the behavior of a person admitted to the hospital to the extent necessary for their medical care or protection. Additionally, the administrator of a psychiatric hospital must comply with the standards established by the Minister of Health, Labour and Welfare regarding the treatment of persons admitted to the hospital (Mental Health and Welfare Act, Articles 36 and 37).

15.3 Respectful Treatment in a Psychiatric Hospital

An administrator of a psychiatric hospital may restrict the activities of a hospitalized person to the extent essential for the medical care and protection of that person (Mental Health and Welfare Act, Article 36).

When isolating a patient for 12 hours or more, as deemed necessary by a designated mental health physician, the following requirements must be met:

(i) Only the patient may enter the isolation room.
(ii) The reasons for isolation, as well as the start and end times, must be recorded in the medical record.
(iii) During the isolation period, careful clinical observation, appropriate treatment, and hygiene must be ensured.
(iv) A physician must examine at least once a day in principle.

Additionally, stricter physical restraints are also subject to standards established by the Minister of Health, Labour and Welfare. Furthermore, even when action restrictions are imposed based on the judgment of a designated physician, restrictions on sending and receiving letters, as well as making telephone calls and visiting administrative agency staff or attorneys acting as agents, are not permitted, as stipulated by the Minister of Health, Labour and Welfare.

15.4 Statistics

Examining the number of patients by type of admission reveals that medical protection admissions account for 50.4%, while voluntary admissions based on the patient's consent account for 48.5%. Regarding the status of psychiatric beds, the number of psychiatric beds in 2022 was 322,000 (as reported in the Medical Facility Survey), and the average length of stay in psychiatric beds was 276.7 days (as reported in the Hospital Report). Although both have decreased and shortened compared to previous years, they remain at very high levels internationally.

Furthermore, when the 274,000 patients hospitalized in psychiatric beds in 2020 are classified by disease category, "schizophrenia, schizophrenia-type disorders, and delusional disorders" account for 141,000, or more than half (Patient Survey).

15.5 Psychiatric Review Boards

Prefectures must establish "Psychiatric Review Boards" to conduct regular reviews of the condition of patients admitted under compulsory measures, as well as to review requests for discharge from hospitalized patients or their families, among other matters. Upon request from patients hospitalized in psychiatric hospitals or their families, the prefectural governor may request the Psychiatric Review Boards to conduct a review.

Suppose the Minister of Health, Labour and Welfare, or the prefectural governor determines that the treatment of a patient hospitalized in a psychiatric hospital violates the provisions of the law or is deemed to be significantly inappropriate, in that case, they may order the administrator of the psychiatric hospital to submit an improvement plan to ensure appropriate treatment or take necessary measures to improve treatment, specifying the matters to be addressed and the deadline for implementation (Mental Health and Welfare Act, Article 38-7) [1].

15.6 Health Centers and Mental Health and Welfare Centers

The prefectural governor shall issue the Certificate of Mental Disorder for Persons with Mental Disabilities to the applicant if the applicant is found to be in a state of mental disability as specified by Cabinet Order (Mental Health and Welfare Act, Article 45).

Prefectures are required to establish Mental Health and Welfare Centers to promote mental health and improve the welfare of persons with mental disabilities by providing complex and difficult counseling and guidance on mental health and welfare, and disseminating knowledge on mental health and welfare (Mental Health and Welfare Act, Article 6), as well as establishing psychiatric hospitals (Mental Health and Welfare Act, Article 19-7).

The primary organization responsible for mental health activities in local communities is "health centers," which conduct mental health and welfare-related activities such as assessing the actual conditions of mental health and welfare within their jurisdiction, providing mental health and welfare consultations, and conducting home visits. Additionally, mental health and welfare centers are institutions that provide technical guidance and assistance to health centers and other local organizations related to mental health. They are established in all prefectures and designated cities, and are staffed with specialized technical personnel, including psychiatrists, mental health and welfare workers, clinical psychologists, and public health nurses.

Furthermore, efforts should be made to establish the position of mental health and welfare counselors within the staff, and the director should be a physician with expertise in mental health and welfare. The duties of mental health and welfare centers include technical guidance and assistance to health centers and other mental health-related institutions, education and training for staff of such institutions, research and surveys, and mental health and welfare counseling (for complex or difficult cases) [1].

References

1. Act on Mental Health and Welfare for Persons with Mental Disorders or Disabilities, Act No. 123 of May 1, 1950, Japanese Law Translation. https://www.japaneselawtranslation.go.jp/ja/laws/view/4235.
2. New Essential Forensic Medicine (6th edition, in Japanese), 2019, Ishiyaku Shuppan, ISBN978-4-263-73189-5.

Afterword

This book serves as an introductory guide to the big picture of Japan's health policies and laws, grounded in the nation's unique culture and history. What we most want to convey through this book is that even Japan's health policies, often praised as highly effective, once faced extremely challenging circumstances.

Today's Japanese health policy is built upon existing systems and laws, with applied systems layered on top. For example, after establishing the personnel supporting health care and the hospitals providing it through the Medical Practitioners Act and the Medical Care Act, the National Health Insurance Act was enacted to achieve universal health coverage (UHC). Understanding "the sequence" in which these systems were established is also considered important when other countries reference Japan's health policy to formulate their own.

Furthermore, the authors of this book include not only professors with long experience in university student education, but also students from both Japan and the United States who contributed to the writing. We aimed to create content that is accessible from a student's perspective upon first reading. Therefore, we prioritized narrative flow over strict legal or terminological precision in this book.

While every effort has been made to ensure accuracy, inherent limitations exist due to differences between the Japanese and English languages, and laws are constantly being amended. The editors and authors of this book are not responsible for the accuracy, reliability, or currency of the legislative material provided on this website, or for any consequence resulting from use of the information in this website. For all purposes of interpreting and applying law to any legal issue or dispute, users should consult the original Japanese texts published in the Official Gazette.

Currently, there is no English text that provides a detailed explanation of Japan's health policy. However, the English version of the summary of the "White Paper on Health, Labour and Welfare," published annually by the Ministry of Health, Labour and Welfare, and the foreign-language legal database created by Japan's Ministry of Justice help to understand Japan's health policy. These two documents also serve as references for this book.

Finally, we would like to express our sincere gratitude to Mr. Badrinath Parthasarathy, Mr. Ben Ingraham, Ms. Saki Kasai, and Ms. Chihiro Haraguchi of Springer Nature for their editorial support of this book.

© The Editor(s) (if applicable) and The Author(s), under exclusive license to Springer Nature Singapore Pte Ltd. 2026

J. Ezaki et al. (eds.), *Health Policy and Law in Japan*,
https://doi.org/10.1007/978-981-95-9241-8

Index

© The Editor(s) (if applicable) and The Author(s), under exclusive license to Springer Nature Singapore Pte Ltd. 2026
J. Ezaki et al. (eds.), *Health Policy and Law in Japan*,
https://doi.org/10.1007/978-981-95-9241-8

MIX
Papier aus verantwortungsvollen Quellen
Paper from responsible sources
FSC® C105338

FSC
www.fsc.org

If you have any concerns about our products,
you can contact us on
ProductSafety@springernature.com

In case Publisher is established outside the EU,
the EU authorized representative is:
Springer Nature Customer Service Center GmbH
Europaplatz 3, 69115 Heidelberg, Germany

Printed by Libri Plureos GmbH
in Hamburg, Germany